Casting Handbook

Casting Handbook

Hannah Wells

NY RESEARCH PRESS

New York

Published by NY Research Press
118-35 Queens Blvd., Suite 400,
Forest Hills, NY 11375, USA
www.nyresearchpress.com

Casting Handbook
Hannah Wells

International Standard Book Number: 978-1-63238-868-1 (Hardback)

Cataloging-in-Publication Data

Casting handbook / Hannah Wells.
    p. cm.
Includes bibliographical references and index.
ISBN 978-1-63238-868-1
1. Metal castings. 2. Founding. I. Wells, Hannah.
TS236 .C37 2022
671.2--dc23

Contents

Preface

The process through which a liquid material is delivered into a mold, which contains a hollow cavity of the intended shape, and then being allowed to cool is known as casting. It is generally used to produce sophisticated shapes which are difficult or uneconomical to make using other methods. Metals and other time setting materials are usually used for casting. It is used in the making of a variety of products such as bronze structures, concrete structures, collectible toys weapons and tools. The traditional methods of casting are lost-wax casting, sand casting and plaster mold casting. The modern casting processes are classified into two categories, expendable and non-expendable casting. The topics included in this book on casting are of utmost significance and bound to provide incredible insights to readers. Some of the diverse topics covered in this book address the varied branches that fall under this category. It will serve as a valuable source of reference for graduate and post graduate students.

A detailed account of the significant topics covered in this book is provided below:

Chapter 1- The process wherein a liquid material is poured into a mold and then allowed to solidify is termed as casting. Some of the common methods of casting are sand casting, shell casting and investment casting. This chapter has been carefully written to provide an introduction to these methods of casting.

Chapter 2- There are numerous types of casting. A few of these are metal casting, continuous casting, die casting, gravity casting, centrifugal casting, permanent mold casting, sand casting and investment casting. The diverse applications of these types of casting have been thoroughly discussed in this chapter.

Chapter 3- There are various properties of different material which have to be considered during the casting process. A few of these are mechanical properties of cast iron, tensile properties of aluminum foundry alloys and the fracture toughness of metal castings. The topics elaborated in this chapter will help in gaining a better perspective about these casting properties.

Chapter 4- There are a wide variety of alloys which are chosen for their diverse physical and mechanical properties. A few of the categories of alloys used for casting are aluminum-silicon alloys, zinc casting alloys and magnesium alloys. This chapter discusses in detail the casting processes related to these alloys.

Chapter 5- There are diverse processes which take place after casting has been completed. A few of these processes are heat treatment, machining of the cast materials, and painting and finishing. These post casting processes have been thoroughly discussed in this chapter.

It gives me an immense pleasure to thank our entire team for their efforts. Finally in the end, I would like to thank my family and colleagues who have been a great source of inspiration and support.

Hannah Wells

An Introduction to Casting

The process wherein a liquid material is poured into a mold and then allowed to solidify is termed as casting. Some of the common methods of casting are sand casting, shell casting and investment casting. This chapter has been carefully written to provide an introduction to these methods of casting.

Casting is defined as a manufacturing process in which molten metal pure into a mold or a cavity of desire shape and allow to solidify which form a predefine shape. This process is widely used to manufacture complex parts which cannot be made by other processes. All major parts like bed of lathe machine, milling machine bed, IC engine component etc. are made by this process.

Working Process

There are many types of casting which works differently but all these processes involve followings steps:

- First metal is melted in a suitable furnace.

- Now molten metal poured into a predefine cavity.

- The molten metal allows to solidify at desire cooling rate.

- Removal of cast part from mould and clean it for further processes like machining, surface finishing polishing etc.

Advantages

Casting has following advantages over other manufacturing process:

- It can create any complex structure economically.

- The size of object doesn't matter for casting.

- The casting objects have high compressive strength.

- All structure made by casting has wide range of properties.

- This can create an accurate object.

- All material can be cast.

- It creates isotropic structure.

- It is cheapest among all manufacturing processes.

- Composite component can be easily made by casting.

Disadvantages

Along these advantages, casting has following disadvantages:

- It gives poor surface finish and mostly requires surface finish operation.

- Casting defects involves in this process.

- It gives low fatigue strength compare to forging.

- It is not economical for mass production.

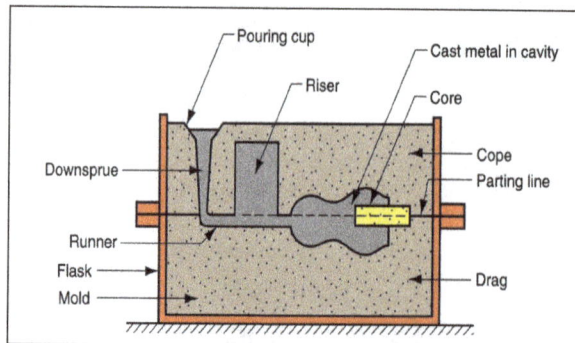

Casting Terminology.

- Flask: The moulding flask is used to hold the sand mould. The sand mould has desire cavity of object which to be casted. The sand is rammed into the flask to create sand mould in which metal is poured to get desire shape. It is created into minimum two pieces which allows removal of pattern easily.

- Cope: The upper part of moulding flask is known as cope.

- Drag: The lower part of moulding flask is known as drag.

- Cheeks: When the moulding flask made into more than two parts, the intermediate parts are known as cheeks. These are used in complex casting.

- Pattern: Pattern is replica of object to be created. It is made by either wood, wax or other suitable material. It is placed into moulding flask and sand rammed over it which forms an object cavity into sand.

- Pouring Basin: It is a funnel shape cavity at the top of the mould. The metal is poured into pouring basin from where it is supplied at different parts of mould.

- Runner: Runner is a horizontal passage of molten metal. It connects sprue to getting system. Normally it is situated at lower half of mould.

- Riser: Riser is used as reservoir of molten metal when pouring of molten metal has stopped. When the cavity is filled by molten metal, the pouring is stopped which allows solidifying object. During solidification, volumetric shrinkage takes place which reduces the desire size and shape of object. The riser is provided into the mould which supplies the molten metal to remove effect of volumetric shrinkage during solidification. These are further divided into top riser, blind riser, side riser etc.

- Sprue: It is a passage which connects pouring basin to the runner. It controls the flow of molten metal from pouring basin. It is tapered in shape.

- Ingate: It is the entry point through which molten metal enters into the actual mould cavity.

- Core: Core is used to cast hollow cavity. It is also a sand structure and placed at right place into mould cavity where hollow part is to be created. The metal poured into mould cavity does not fill the part at which core is placed thus form a hollow cavity.

- Chaplets: These are supporting components of core. These used to support and hold the core into mould cavity. These protect the core from various forces encounter in casting.

- Chills: Chills are generally solid metal pieces which are placed into cavity to increase cooling rate. Mainly it is used to create direction solidification of molten metal. They have high thermal conductivity.

- Vents: These are small passages made in mould which allow to escape the gases during solidification.

Classification of Casting Processes

Casting is widely used any many different shapes and material can be cast by it. There are various method of casting available which are used for different shapes and material. Casting has following types:

Sand Casting

It is widely used for casting different process. Sand is easily available and has high refectory property so it is used in casting. It is done by following steps:

- Design is made by using software or manually.

- A wooden pattern is created in pattern shop. Generally patterns are made into two half and can be increased accordingly to complexity.

- The pattern is placed into the flask and mixture of sand and clay with water pour into it. The runner, riser, core, gating system is also fit into it.

- When the mould gets hard the pattern is removed from mould and molten metal pour into it.

- The metal is allowed to get solidify into the casting.

- After solidification cast is removed from casting and send to machine shop for machining.

- The sand casting is used for all metal and at low cost. An another advantage is that it can be used for very complex shape. It gives poor surface finish.

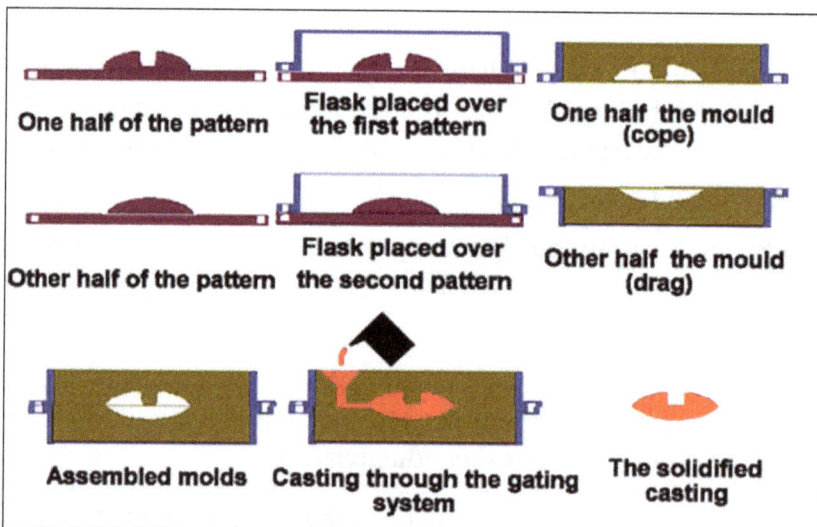

One half of the pattern

Flask placed over the first pattern

One half the mould (cope)

Other half of the pattern

Flask placed over the second pattern

Other half the mould (drag)

Assembled molds

Casting through the gating system

The solidified casting

Shell casting.

It is an another method of casting and used to cast thin section like turbine blade etc. This casting is different from sand casting. A metallic pattern is used in this type of casting. It consists following steps:

- First a metallic pattern is cast. The metal is used for casting is generally aluminum or cast iron.

- The patterns is heated up to 250 degree and put into flask.

- The flask is filled with sand resin mixture. The resin gets solidify immediately after gating contact with heated pattern.

- After the sand solidify the pattern and the extra sand taken out. Now a shell of cavity is created. This shell is further heated into burner which allow proper bond.

- The metal is poured into the shell and allows to solidify.

- After solidification cast is removed from shell and send for machining.

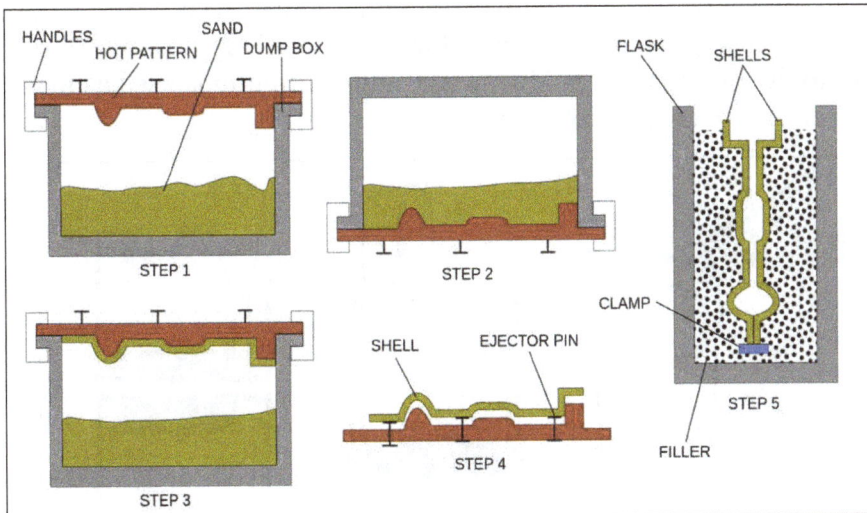

Shell casting.

Investment Casting

In this type of process, wax pattern is used. The pattern is first created by wax dipped into refectory material. This refectory material make as shall outside the wax pattern. After it mould is heated which allow waxing out from mould. Now the molten metal poured into cavity formed by it and allows solidifying. The cast is taken out after proper solidification of cast and send for machining.

The main advantage of this process is that a very high accuracy and surface finish can obtain by it. It is used for complex shape and batch production.

Investment casting.

Plaster Mould Casting

These method uses plaster mould for casting. First plaster mould is created using patterns. After removal of pattern, the plaster mould allows to dry into an oven. After
dried, molten metal is poured into it and allow solidifying. After complete solidification
cast is sand to the machine shop. Mostly zinc and aluminum is used as molten metal.
This casting is used to create prototype.

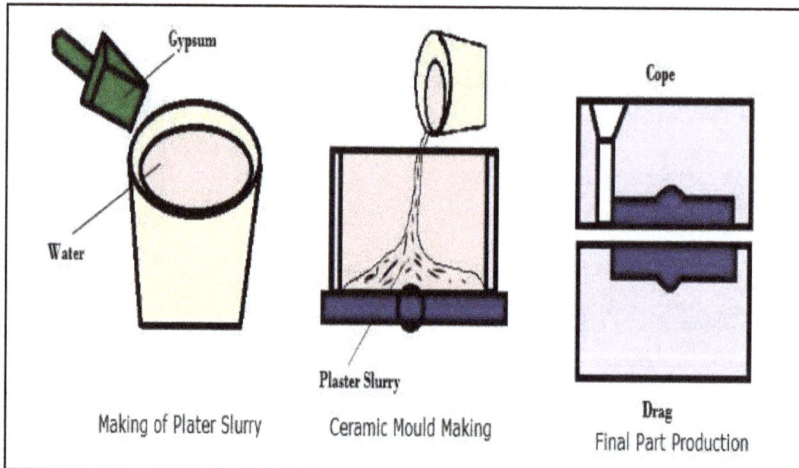

Plaster mould casting.

Ceramic Mold Casting

Ceramic mold casting is same as investment casting but it does not use wax pattern.
The slurry of ceramic and liquid binder is pours on pattern which is easily solidify.
There is no wax pattern is used so there is no limitation of size of casting. This type
of casting is mainly used to cast heavy component like valve bodies, military equipment etc.

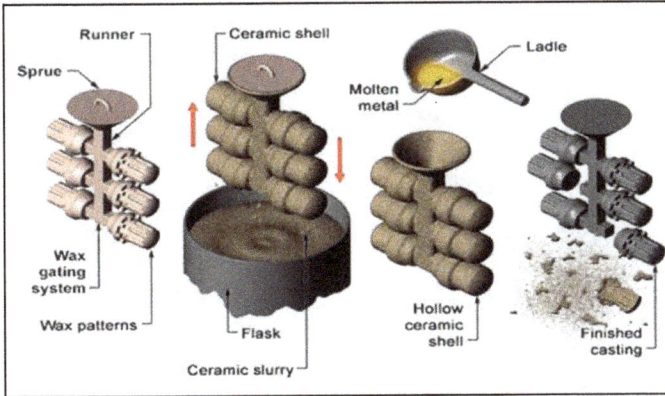

Ceramic mold casting.

Pressure Dies Casting

This casting is mostly used to cast aluminum, lead etc. In this casting a liquid metal is pumped at high pressure into a metallic die and allow to solidify. The metal is allowed to solidify there and take out after solidification .The die is again used for next cast. It is well suitable for batch production of casting low melting point metal. It is not suitable for high melting temperature metals. The tooling cost is also high.

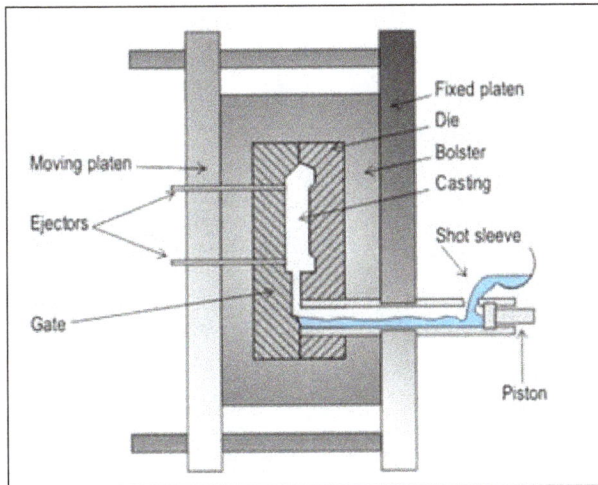

Pressure die casting.

Centrifugal Casting

It is one of the most suitable casting processes of casting symmetrical cylindrical component. In this process a liquid metal is poured at the center of a rotating die. The die rotate and a centrifugal force act on the molten metal which forces it to towards circumference. It is used to create hollow shape. The light impurities crowed near center which is removed by machine. This process eliminates the use for core and gating system. This type of casting is used to make pipes etc.

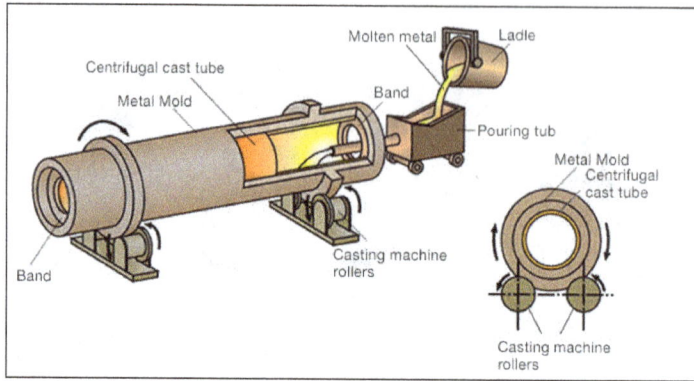

Centrifugal casting.

Continuous Casting

It is a different casting process which is used to create continuous cast. In this process we do not use mould or cope and drag. It is different in principle. In this process the molten metal poured into a trash which is connected to a copper pipe. The copper pipe is surrounded by water cooling. The metal is directly or immediately cooled after when running through pipe. The casting product takes out from other side. This process continuously run and molten metal continuously pour into it. It is used to create square or other shape simple block which further used for rolling or other process.

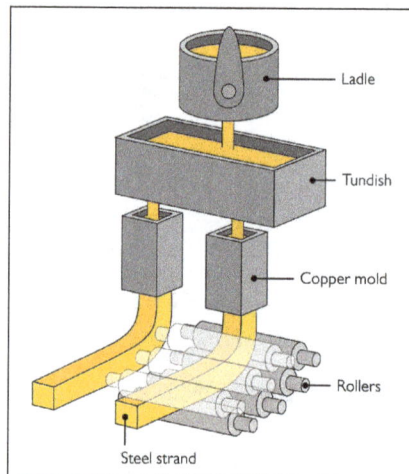

Continuous casting.

Major Casting Components

Ingenuity and new technological advancements are transforming todays' metal casting industry that has become innovative, high-tech, challenging, clean, and safe. Multiple processes have been developed in the industry where each process is specific to the metal used and the results desired. Within each process there are several variables that impact the design of final product. Today, castings are used in a number of markets in a variety of applications that range from manufacturing to home decor.

The diverse range of casting alloys and their flexibility offers the selection of most economical materials to fulfill the prerequisites of a specific application. Every single alloy has particular physical and mechanical characteristics, as well as its own casting properties, machinability, weldability, corrosion resistance, heat treatment properties, and other characteristics.

The metal casting process today has become integral to manufacturing industry and can be used to create complex geometric parts with relative ease, irrespective of the size of part. Cast metal products are found in 90% of manufactured goods and equipment including critical components for aircraft and automobiles to home appliances and surgical equipment. Also, the process is very economical and generates little waste, which can be reheated and used again.

Some of the major casting components are:

- Automobile Casting Equipment.

- Pumps and Valves Component.

- Scientific Casting Equipment.

- Aluminum & all other Non-Ferrous Parts.

- Chemical Process Equipment.

- Earthmoving Equipment.

- Agriculture Equipment.

The table given below shows the the different types of casting components and their share:

End-Use Markets	Percent of Castings Shipped
Automotive and Light Truck	35%
Pipe and Fitting	15%
Construction, Mining & Oil Field Equipment	6%
Internal Combustion Engines	5%
Railroad	5%
Valves	5%
Farm Machinery	3%
Municipal Casting (manhole covers, grates etc.)	3%
Pumps & Compressors	2%
Other Markets	21%

1. Automobile Casting Equipment: The automobile industry is a major market segment for cast products using ferrous, non-ferrous metals and its alloys. Different casting techniques and methods are used to design automobile components, which are light in weight, easy to maneuver and economical.

Zinc die casting and aluminum die casting are two of the most popular methods that are used to create cast components for automobiles.

2. Pumps and Valves Components: Metal casting is integral to the manufacturing of valves and pumps as it lends itself readily to design the complex geometries of valves and component. Sophisticated casting techniques have been developed that can be used to leverage the advantage of very close tolerances and consistent quality.

Valves and pumps can be made either by using ferrous or non-ferrous metal casting techniques. Examples of ferrous casting include malleable iron casting, gray iron casting and ductile iron casting. Some of the common methods of non-ferrous casting are – silicon brass casting, red brass casting, tin brass alloy casting, and more.

3. Aircraft Components: Metal casting has long been favored for designing aircraft components. Casting methods are used to design and produce components of complicate geometries and shapes for civilian aircrafts, jet fighters and helicopters. Magnesium and aluminum metal along with their alloys present a variety of choices. This is because of the fact that these metals are light in weight, low density, possess excellent high temperature mechanical properties, good corrosion resistance and can function under highly demanding conditions.

Some of the aircraft components that the metal casting is used to manufacture are – engines, airframes, skins, fasteners, compressor blades, turbine disks, helicopter transmission casings, auxiliary gearboxes, generator housings, intermediate compressor casings for turbine engines.

4. Cast Iron Components: Known for their superior strength and long service life, cast iron components have emerged as the most preferred components for manufacturing industry. Cast iron components can also withstand high wear and tear and hence can be used in the most demanding conditions.

Grey iron with flake graphite and ductile iron with spheroidal graphite are the most popular methods to produce cast iron components.

Some of the important cast iron components for manufacturing industry include - fly wheels, machine bases, engine blocks, piston rings, brake, discs and drums; heavy duty gears, pistons, rolls for rolling mills, gear case, valves, tubes and door hinges; parts for heavy trucks, farm and earth moving equipment like axle journals, gear drives, crankshafts, pull hooks and wheels.

5. Scientific Casting Equipment: Because of its flexibility, ease of use and durability, metal casting is widely preferred across the world to design and create scientific equipment. A variety of alloy choices are available, where each alloy has its own mechanical and physical properties along with distinct casting characteristics that offer the selection of most economical materials to meet a particular application.

Some of the important scientific casting equipment are – test and measurement equipment, microscopes, centrifuges, autoclaves, imaging optics, and other lab equipment.

6. Chemical Process Equipment: Metal casting is one of the most preferred methods for producing chemical process equipment. Characteristics such as light weight and strength of cast components make them an ideal choice for chemical process equipment. Corrosion resistance is a major factor that is kept in mind before selecting a metal for this purpose. Titanium for anti-corrosive properties is preferred for making chemical process equipment.

Some of the examples of chemical process equipment are – production equipment that transports corrosive materials; manufacturing equipment, such as – reflux towers, filters and pressure vessels; heating coils in laboratory autoclaves and heat exchangers.

7. Earthmoving Equipment: Because of their superior strength, design flexibility and ease of use of cast components, metal casting is preferred for manufacturing earthmoving equipment. Some of the metal casting parts that are used in earthmoving equipment are - axle journals, gear drives, crankshafts, pull hooks and wheels.

8. Agriculture Equipment: Agricultural equipment enhance productivity and reduce drudgery, however, these equipment also have to withstand extreme conditions and heavy wear and tear. Practically, this equipment must also be affordable to the

farmers and profitable for the manufacturers. This aspect of manufacturing productivity and efficiency of manufacturing equipment makes metal casting an integral process to this industry.

Metal casting is used to design and produce an extensive variety of agricultural equipment, including – measuring instruments, irrigation equipment, harvesting machines, fencing equipment, grain processing machines, water pumps, and more.

Aluminum and other Non-ferrous Parts

Aluminum castings have always played a key role in the development of aluminum industry since its inception in the 19th century. The earlier aluminum products designed for commercial reasons were castings, such as cooking utensils and decorative parts that utilized the novelty and utility of this new metal. Those starting applications quickly expanded to treat the requirements of an extensive range of engineering specifications.

Different Types of Pattern Allowance in Casting

The pattern is replica of casting but it has slightly large dimension. This change in dimensions in pattern and casting are due to various allowance used in casting. When the cast solidify, it shrinks at some extent due to metal shrinkage property during cooling, so a pattern make slightly larger to compensate it. There are other reasons which are due to poor surface finish and casting limitations, casting make slightly larger so it can machine or polish further. So the pattern hence cavity for casting forms slightly larger which can compensate all these drawbacks of casting. The change in dimensions of pattern and casting are known as allowance.

Types of Pattern Allowance

Allowance can be classified into following types:

1. Shrinkage allowance: Shrinkage is defined as reduce the dimension of casting during solidification or during cooling. This is general property of all materials. Some metal shrinks more, some less but every material shrinks. There are three types of shrinkage.

- Liquid Shrinkage,

- Solidification Shrinkage,

- Solid Shrinkage.

The liquid shrinkage and solidification shrinkage are compensated by suitable riser but solid shrinkage does not compensated by it so the pattern is made slightly larger to compensate shrinkage. This is known as shrinkage allowance.

2. Draft Allowance: When the pattern is removed from mould, the parallel surface to the direction at which pattern is withdrawn, damaged slightly and convert into slightly tapered surfaces. To compensate these changes, these parallel surfaces on patterns are made slightly tapered (1-2 degree). This allows easy removal of pattern from mould and does not affect the actual dimension of casting. These are known as draft allowance.

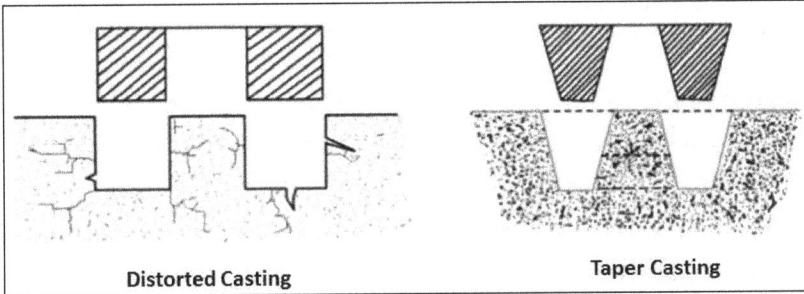

Draft Allowance.

3. Machining Allowance: As we known the casting gives poor surface finish and rough surfaces. Also, the Pattern is made manually which does not fixed accurate angles and dimensions of mating parts. But we need a proper finished and dimensionally accurate casting. To avoid these problems, casting made slightly larger and after solidification machining is done on it for better surface finish and accurate dimensions. This change in dimension of casting is known as machining allowance.

4. Distortion Allowance: When casting of very thin surfaces like V Shape, U shape etc. it will distort or damage during pattern removal or during casting. To avoid this problem, a chamber is provided on pattern to compensate distortion during pattern removal. This change in casting dimension is known as distortion allowance.

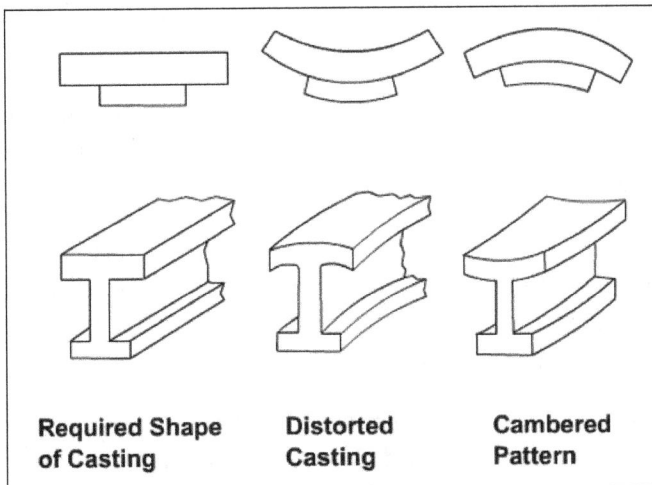

Distortion Allowance.

5. Rapping Allowance: When the pattern is removed from casting, it will slightly increase the dimension of casting. So to compensate these changes, the pattern is made slightly smaller from casting. This change in dimension is known as rapping allowance.

Allowances Commonly Provided on a Pattern

1. Draft Allowances: A tapper of about 1 percent is provided to all surfaces perpendicular to the parting line to facilitate the easy removal without damaging the vertical sides of the mould. This is known as the draft allowances. Taper may be expressed in millimeters per meter or in degrees.

Wooden patterns require more taper than metallic patterns. The taper may varies from 1 to 2 degrees. For pockets or deep cavities this values may higher and for large castings it may be reduces to less than 1/2 degree. The draft allowance is shown in figure.

2. Shrinkage Allowance: Almost all metals shrink during solidification and contract with further cooling to room temperature. To compensate this effect, a pattern is made slightly larger than the actual dimensions of the finished casting.

The value of shrinkage allowance depends upon the metal to be cast and, to some extent, on the nature of the casting. The shrinkage allowance is shown in figure.

The Shrinkage allowance is usually taken as 1 percent for cast iron, 2 percent for steel, 1.5 percent for aluminum, 1.5 percent for magnesium, 1.6 percent for brass and 2 percent for bronze. In order to eliminate the needs for recalculating all the dimensions of a casting.

Pattern makers use a shrink rule or Contraction scale. It is longer than the standard 1-Foot rule; its length differs for the different metals of the casting.

3. Machining Allowance: The dimensions of a pattern are oversized than the actual casting required. This is because, the layer of metal that is removed by machining to obtain better surface finish. The ferrous metal requires more machining allowance than non-ferrous metals.

The amount of machining allowance depends upon certain factors like shape, size, type of metal to be casted, method of machining, degree of surface finish required, etc. The amount of this allowances is usually varies from 1.5 to 15 mm. The machining allowance is shown in figure.

4. Distortion Allowance: Sometimes, intricately shaped or irregular casting distort during solidification. In such cases, it is necessary to distort the pattern intentionally to obtain a casting with the desired shape and dimension.

The distortion is caused by non-uniform solidification or cooling of metal at thin and thick portions of the part. I-section and U-section beams having different top and bottom flange thickness suffers with distortion. This is shown in figure.

5. Shake Allowance: The shake allowance is also known as rapping allowance. The pattern when withdrawn from the mould it distort the sides and shape of the cavity.

To avoid this, the pattern is shaked to create a small void or gap between the mould and pattern surface for easy removal. This increases the size of cavity and hence to compensate this, the size of the pattern is slightly smaller than castings. Shake allowance is considered as negative allowance and is shown in figure.

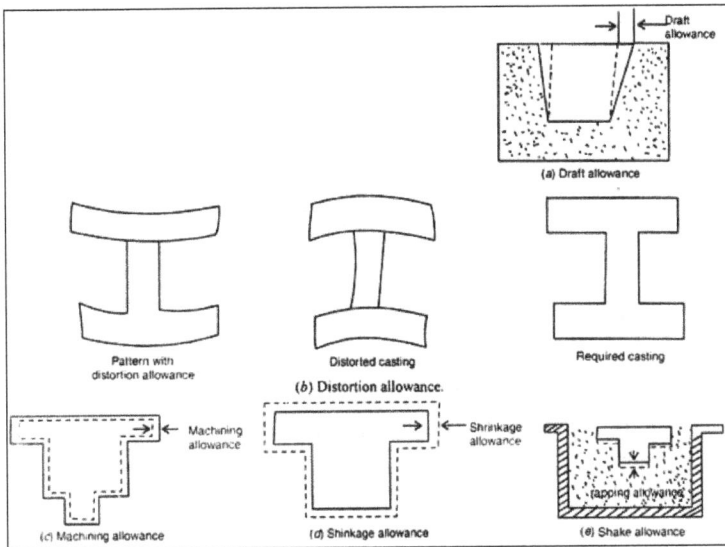

Pattern Allowance.

Metal Casting Operation

Pouring of the Metal

When manufacturing by metal casting, pouring refers to the process by which the molten metal is delivered into the mold. It involves its flow through the gating system and into the main cavity (casting itself).

Goal: Metal must flow into all regions of the mold, particularly the casting's main cavity, before solidifying.

Factors of Pouring

1. Pouring Temperature: Pouring temperature refers to the initial temperature of the molten metal used for the casting as it is poured into the mold. This temperature will obviously be higher than the solidification temperature of the metal. The difference between the solidification temperature and the pouring temperature of the metal is called the superheat.

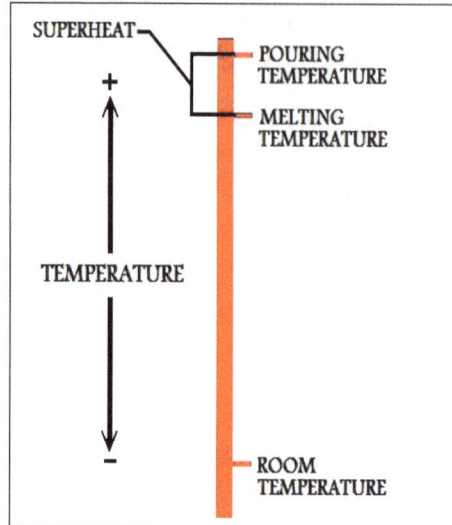

2. Pouring Rate: Volumetric rate in which the liquid metal is introduced into the mold. Pouring rate needs to be carefully controlled during the metal casting operation, since it has certain effects on the manufacture of the part. If the pouring rate is too fast, then turbulence can result. If it is too slow, the metal may begin to solidify before filling the mold.

3. Turbulence: Turbulence is inconsistent and irregular variations in the speed and direction of flow throughout the liquid metal as it travels though the casting. The random impacts caused by turbulence, amplified by the high density of liquid metal, can cause mold erosion. An undesirable effect in the manufacturing process of metal casting, mold erosion is the wearing away of the internal surface of the mold. It is particularly detrimental if it occurs in the main cavity, since this will change the shape of the casting itself. Turbulence is also bad because it can increase the formation of metal oxides which may become entrapped, creating porosity in the solid casting.

Fluidity

Pouring is a key element in the manufacturing process of metal casting and the main goal of pouring is to get metal to flow into all regions of the mold before solidifying. The properties of the melt in a casting process are very important. The ability of a particular casting melt to flow into a mold before freezing is crucial in the consideration of metal casting techniques. This ability is termed the liquid metals fluidity.

Test for Fluidity

In manufacturing practice, the relative fluidity of a certain metal casting melt can be quantified by the use of a spiral mold. The geometry of the spiral mold acts to limit the flow of liquid metal through the length of its spiral cavity. The more fluidity possessed by the molten metal, the farther into the spiral it will be able to flow before hardening.

The maximum point the metal reaches upon the casting's solidification may be indexed as that melts relative fluidity.

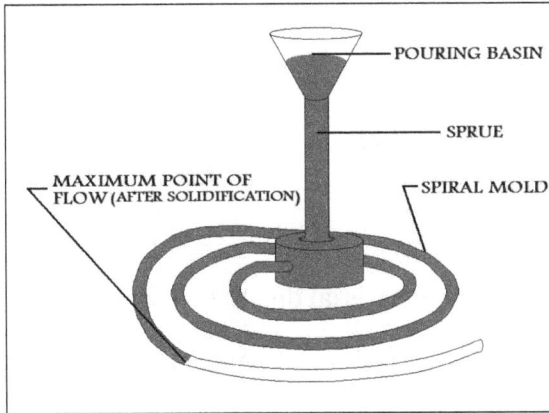

Spiral mold test.

How to Increase Fluidity in Metal Casting

- Increase the superheat: If a melt is at a higher temperature relative to its freezing point, it will remain in the liquid state longer throughout the metal casting operation, and hence its fluidity will increase. However, there are disadvantages to manufacturing a metal casting with an increased superheat. It will increase the melts likelihood to saturate gases, and the formation of oxides. It will also increase the molten metals ability to penetrate into the surface of the mold material.

- Choose a eutectic alloy, or pure metal: When selecting a manufacturing material, consider that metals that freeze at a constant temperature have a higher fluidity. Since most alloys freeze over a temperature range, they will develop solid portions that will interfere with the flow of the still liquid portions, as the freezing of the metal casting occurs.

- Choose a metal with a higher heat of fusion: Heat of fusion is the amount of energy involved in the liquid-solid phase change. With a higher heat of fusion, the solidification of the metal casting will take longer and fluidity will be increased.

Shrinkage

Most materials are less dense in their liquid state than in their solid state and denser at lower temperatures in general. Due to this nature, a metal casting undergoing solidification will tend to decrease in volume. During the manufacture of a part by casting this decrease in volume is termed shrinkage. Shrinkage of the casting metal occurs in three stages:

- Decreased volume of the liquid as it goes from the pouring temperature to the freezing temperature.

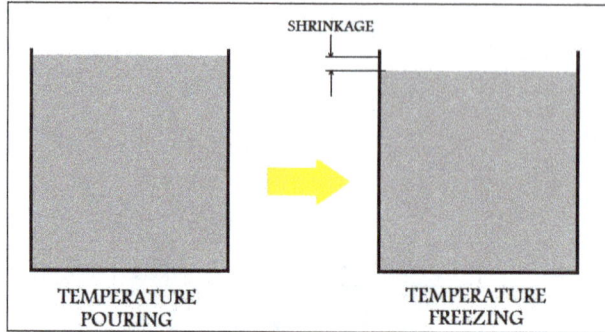

SHRINKAGE

TEMPERATURE
POURING

TEMPERATURE
FREEZING

- Decreased volume of the material due to solidification.

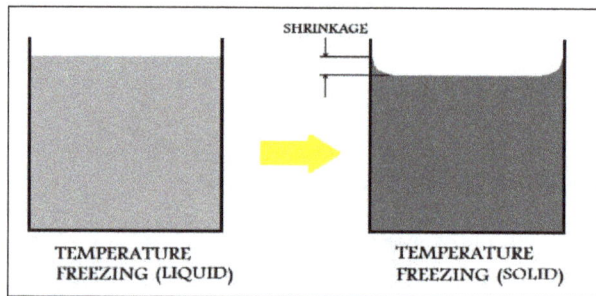

SHRINKAGE

TEMPERATURE
FREEZING (LIQUID)

TEMPERATURE
FREEZING (SOLID)

- Decreased volume of the material as it goes from freezing temperature to room temperature.

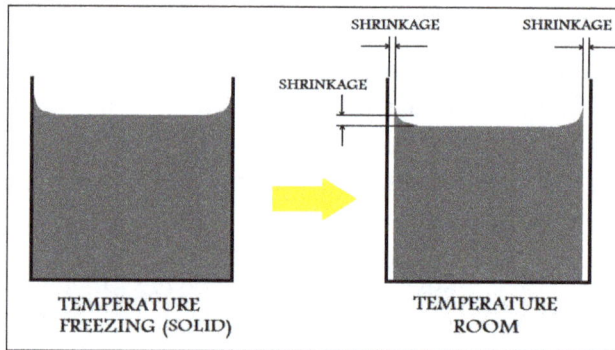

SHRINKAGE   SHRINKAGE

SHRINKAGE

TEMPERATURE
FREEZING (SOLID)

TEMPERATURE
ROOM

Risers

When designing a setup for manufacturing a part by metal casting, risers are almost always employed. As the metal casting begins to experience shrinkage, the mold will need additional material to compensate for the decrease in volume. This can be accomplished by the employment of risers. Risers are an important component in the casting's gating system. Risers, (sometimes called feeders), serve to contain additional molten metal. During the metal's solidification process, these reservoirs feed extra material into the casting as shrinkage is occurring. Thus, supplying it with an adequate amount of liquid metal. A successful riser will remain molten until after the metal casting solidifies. In order to reduce premature solidification of sections within the riser, in many manufacturing operations, the tops of open risers may be covered with an insulating compound, (such as a refractory ceramic), or an exothermic mixture.

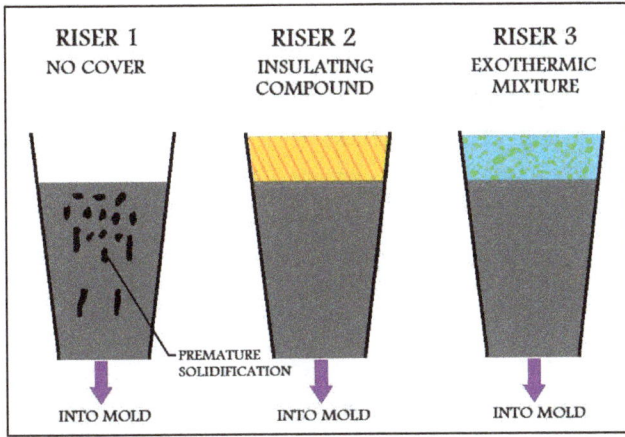

Porosity

One of the biggest problems caused by shrinkage, during the manufacture of a cast part, is porosity. It happens at different sites within the material, when liquid metal cannot reach sections of the metal casting where solidification is occurring. As the isolated liquid metal shrinks, a porous or vacant region develops.

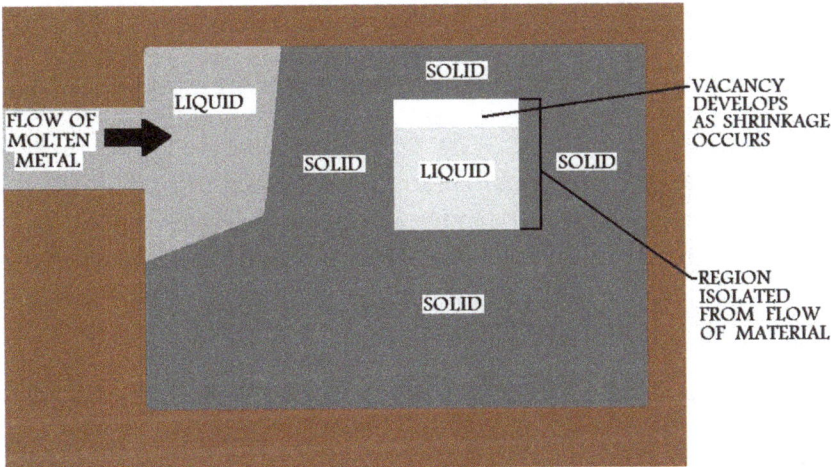

Development of these regions can be prevented during the manufacturing operation, by strategically planning the flow of the liquid metal into the casting through good mold design, and by the employment of directional solidification. Note that gases trapped within the molten metal can also be a cause of porosity. Although proper metal casting methods can help mitigate the effects of shrinkage, some shrinkage, (like that which occurs in the cooling of the work metal from the top of the solid state to room temperature), cannot be avoided. Therefore, the impression from which the metal casting is made is calculated oversized to the actual part, and the thermal expansion properties of the material used to manufacture the part will be necessary to include in the calculation.

Other Defects

The formation of vacancies within the work material due to shrinkage is a primary concern in the metal casting process. There are numerous other defects that may occur, falling into various categories:

- Metal Projections: The category of metal projections includes all unwanted material projected from the surface of the part, (ie. fins, flash, swells, ect.). The projections could be small, creating rough surfaces on the manufactured part, or be gross protrusions.

- Cavities: Any cavities in the material, angular or rounded, internal or exposed, fit into this category. Cavities as a defect of metal casting shrinkage or gases would be included here.

- Discontinuities: Cracks, tearing, and cold shuts in the part qualify as discontinuities. Tearing occurs when the metal casting is unable to shrink naturally and a point of high tensile stress is formed. This could occur, for example, in a thin wall connecting two heavy sections. Cold shuts happen when two relatively cold streams of molten metal meet in the pouring of the casting. The surface at the location where they meet does not fuse together completely resulting in a cold shut.

- Defective Surface: Defects affecting the surface of the manufactured part. Blows, scabs, laps, folds, scars, blisters, etc.

- Incomplete Casting: Sections of the metal casting did not form. In a manufacturing process causes for incomplete metal castings could be; insufficient amount of material poured, loss of metal from mold, insufficient fluidity in molten material, cross section within casting's mold cavity is too small, pouring was done too slowly, or pouring temperature was too low.

- Incorrect Dimensions or Shape: The metal casting is geometrically incorrect. This could be due to unpredicted contractions in the part during solidification. A warped casting. Shrinkage of the metal casting may have been miscalculated. There may have been problems with the manufacture of the pattern.

- Inclusions: Unwanted particles contained within the material act as stress raisers, compromising the casting's strength. During the manufacturing process, interaction of the molten metal with the environment, such as the mold surfaces and the outside atmosphere, (chemical reactions with oxygen in particular), can cause inclusions within a metal casting. As with most casting defects, good mold maintenance and process design is important in their control.

Metal Casting Design

Gating System and Mold Design

When selecting to manufacture a part by casting one must consider the material properties and possible defects that this manufacturing process produces. The primary way to control metal casting defects is through good mold design considerations in the creation of the casting's mold and gating system. The key is to design a system that promotes directional solidification. Directional solidification, in casting manufacture, means that the material will solidify in a manner that we plan, usually as uniformly as possible with the areas farthest away from the supply of molten metal solidifying first and then progressing towards the risers. The solidification of the casting must be such that there are never any solid areas that will cut off the flow of liquid material to unsolidified areas creating isolated regions that result in vacancies within the casting's material.

It is important to create an effective manufacturing process. Gating system design is crucial in controlling the rate and turbulence in the molten metal being poured, the flow of liquid metal through the gating system, and the temperature gradient within the metal casting. Hence a good gating system will create directional solidification throughout the casting, since the flow of molten material and temperature gradient will determine how the metal casting solidifies.

When designing a mold for a metal casting or trying to fix or improve upon and existing design you may want to consider the following areas:

1. Insure that you have adequate material: This may seem very obvious, but in the manufacturing of parts many incomplete castings have been a result of insufficient material. Make sure that that you calculate for the volume of all the areas of your casting, accounting for shrinkage.

2. Consider the Superheat: Increasing the superheat, (temperature difference between the metal at pouring and freezing), as mentioned previously can increase fluidity of the material for the casting, which can assist with its flow into the mold. This causes a compromise to the manufacturing process. Increasing the superheat has problems associated with it, such as increased gas porosity, increased oxide formation, and mold penetration.

3. Insulate Risers: Since the riser is the reservoir of molten material for the casting, it should be last to solidify. Insulating the top as mentioned earlier, shown in figure, will greatly reduce cooling in the risers from the steep temperature gradient between the liquid metal of the casting, and the room temperature air.

4. Consider V/A Ratios: In casting manufacture, V/A ratio stands for volume to surface

area or mathematically (volume/surface area). When solidification of a casting begins a thin skin of solid metal is first formed on the surface between the casting and the mold wall. As solidification continues the thickness of this skin increases towards the center of the liquid mass. Sections in the casting with low volume to surface area will solidify faster than sections with higher volume to surface area. When manufacturing a part by metal casting consideration of the V/A ratios is critical in avoiding premature solidification of the casting and the formation of vacancies.

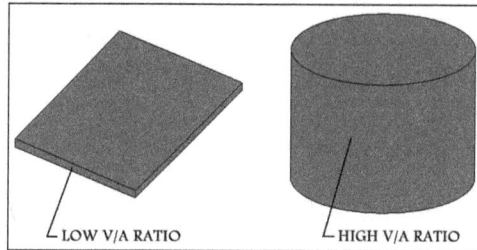

5. Heat Masses: Avoid large heat masses in locations distant to risers. Instead, locating sections of the casting with low V/A ratios further away from the risers will insure a smooth solidification of the casting.

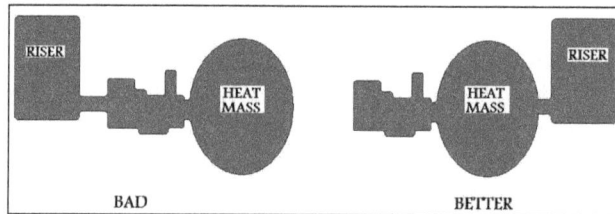

6. Sections of the Casting: The flow of material is very important to the manufacturing process. Do not feed a heavy section through a lighter one.

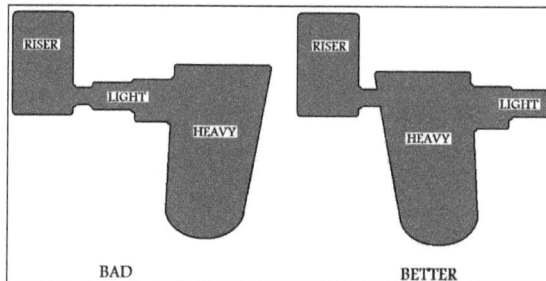

7. Be Careful with Consideration to L, T, V, Y and + junctions: Due to the nature of the geometry of these sections it is likely that they will contain an area where the metal casting's solidification is slower than the rest of the junction. These hot spots are circled in red in figure. They are located such that the material around them, which will undergo solidification first, will cut off the hot spots from the flow of molten metal. The flow of casting material must be carefully considered when manufacturing such junctions. If there is some flexibility in the design of the metal casting and it is possible you may

want to think about redesigning the junction. Some possible design alternatives are shown in figure. These should reduce the likelihood of the formation of hot spots.

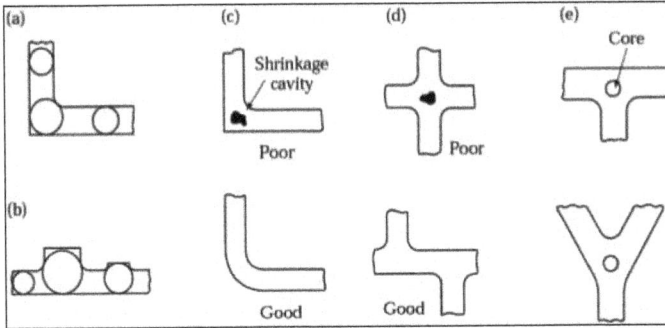

8. Prevent Planes of Weakness: When metal castings solidify, columnar grain structures tend to develop, in the material, pointing towards the center. Due to this nature, sharp corners in the casting may develop a plane of weakness. By rounding the edges of sharp corners this can be prevented.

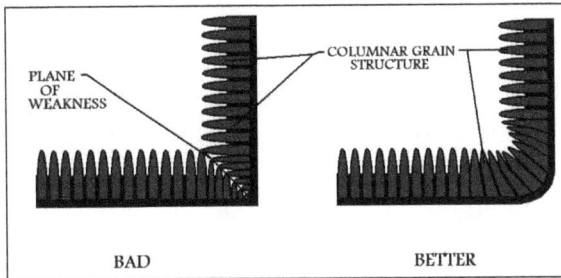

9. Reduce Turbulence: When manufacturing a metal casting, turbulence is always a factor in our flow of molten metal. Turbulence is bad because it can trap gases in the casting material and cause mold erosion. Although not altogether preventable in the manufacturing process, turbulence can be reduced by the design of a gating system that promotes a more laminar flow of the liquid metal. Sharp corners and abrupt changes in sections within the metal casting can be a leading cause of turbulence.

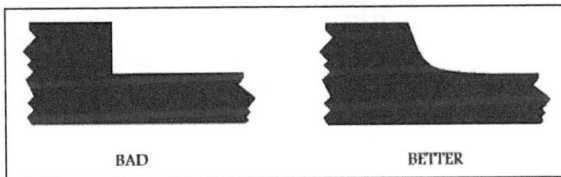

10. Connection between Riser and Casting Must Stay Open: Riser design is very important in metal casting manufacture. If the passage linking the riser to the metal casting solidifies before the casting, the flow of molten metal to the casting will be blocked and the riser will cease to serve its function. If the connection has a larger cross sectional area it will decrease its time to freeze. Good manufacturing design, however, dictates that that we minimize this cross section as much as possible to reduce the waste of

material in the casting process. By making the passageway short we can keep the metal in its liquid state longer since it will be receiving more heat transfer from both the riser and the casting.

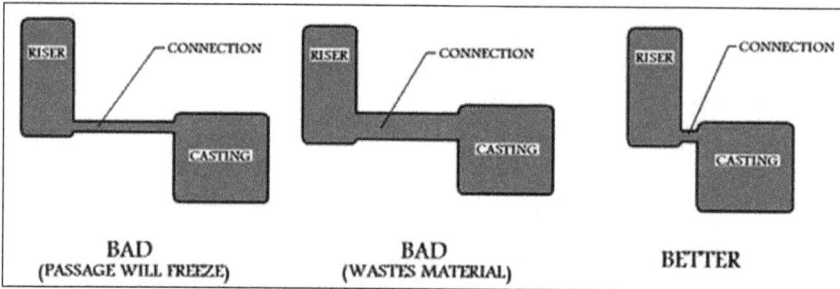

BAD
(PASSAGE WILL FREEZE)

BAD
(WASTES MATERIAL)

BETTER

11. Tapered Down Sprue: Flow considerations for our metal casting manufacture begin as soon as the molten metal enters the mold. The liquid metal for the casting travels from the pouring basin through the down sprue. As it goes downward it will pick up speed, and thus it will have a tendency to separate from the walls of the mold. The down sprue must be tapered such that continuity of the fluid flow is maintained. Remember the fluid mechanics equation for continuity $A_1V_1 = A_2V_2$.

Where V is the velocity of the liquid and A is the cross sectional area that it is traveling through. If you are casting for a hobby and just cannot make these measurements of making A_2 smaller, provided your pouring rate does not become too slow. In other words taper a little more and just adjust your pouring of the casting so that you keep a consistent flow of liquid metal.

12. Ingate Design: The ingate is another aspect of manufacturing design that relates to the flow of metal through the casting's system. The ingate is basically where the casting material enters the actual mold cavity. It is a crucial element, and all other factors of the metal casting's mold design are dependent on it. In the location next to the sprue base the cross sectional area of the ingate is reduced (choke area). The cross sectional reduction must be carefully calculated. The flow rate of casting material into the mold can be controlled accurately in this way. The flow rate of the casting metal must be high enough to avoid any premature solidification. However, you want to be certain that the flow of molten material into the mold does not exceed the rate of delivery into the pouring basin and thus ensure that the casting's gating system stays full of metal throughout the manufacturing process.

Other Flow Considerations

In the manufacturing design phase, when planning the metal casting process, the analysis of the path of flow of liquid metal within the mold must be carefully calculated. At no point in the filling of the casting cavity should two separate streams of liquid metal meet. The result could be an incomplete fusion of the casting material (cold shut), as covered in the defects section under discontinuities.

Use of Chills

Directional solidification is very important to the manufacture of a part during the metal casting process, in order to ensure that no area of the casting is cut off from the flow of liquid material before it solidifies. To achieve directional solidification within the metal casting, it is important to control the flow of fluid material and the solidification rate of the different areas of the metal casting. With respect to the solidification of the metal casting's different sections, regulation of thermal gradients is the key.

Sometimes we may have an area of the metal casting that will need to solidify at a faster rate in order to ensure that directional solidification occurs properly. Manufacture planning and design of flow and section locations within the mold may not be sufficient. Chills act as heat sinks, increasing the cooling rate in the vicinity where they are placed.

Chills are solid geometric shapes of material, manufactured for this purpose. They are placed inside the mold cavity before pouring. Chills are of two basic types. Internal chills are located inside the mold cavity and are usually made of the same material as the casting. When the metal solidifies the internal chills are fused into the metal casting itself. External chills are located just outside of the casting. External chills are made of a material that can remove heat from the metal casting faster than the surrounding mold material. Possible materials for external chills include iron, copper, and graphite. figure demonstrates the use of the two types of chills to solve the hot spot problem in a + and T junction.

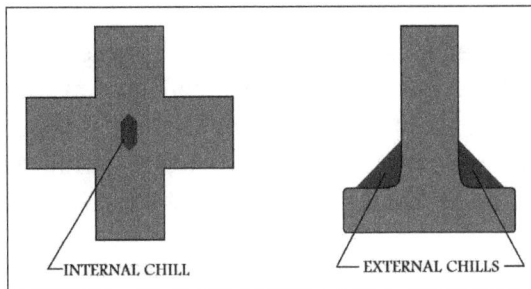

INTERNAL CHILL   EXTERNAL CHILLS

Types of Casting Defects

Some of the casting defects are:

1. Surface Roughness: Too coarse a moulding sand or when pouring temperature is high leads to rough or pebbly surfaces on castings. In steel castings, roughness is produced due to occurrence of surface reaction at mould-metal interface in which iron is oxidised and iron oxide reacts with silica to forms rough compound. Surface reactions sometimes also cause sub-surface porosity or pin holes.

2. Scabs or Buckles: These defects occur due to some sand shearing from the cope surface and as a result there being a layer of metal separated from the casting proper by a layer of sand. Scabs are relatively small particles and buckles are big defects.

These occur due to use of too fine a sand, low permeability, high moisture, uneven ramming of mould, low or intermittent running of molten metal over the sand surface. These can be avoided by using sand with high hot plasticity or low expansion characteristics, using expansion buffer in sand, and rapidly filling the mould.

3. Blow-Holes: They take the form of internal voids (smooth, round or oval holes with a shiny surface), dispersed internal porosity or surface depression as a result of excessive gaseous materials that cannot escape.

They are caused by hard ramming, excessive moisture, low permeability, excessive fine grains, incomplete or improper venting, low temperature of mould and excessive carbonaceous or other organic materials (gas producing ingredients). It is sometimes caused by bad patterns and core-box arrangements, which lead to trapping of gases in blind places in the mould. They can be controlled by taking care of the above points.

4. Pinholes: Surface reactions sometimes cause subsurface porosity or pinholes. In aluminium alloys containing more than about 1% magnesium-magnesium react with water vapour of the mould to form H_2 gas resulting in hydrogen filled pinholes at the surface. In steel castings, subsurface pinholes may result from incomplete de-oxidation of the molten metal.

5. Sand Spots: These appear as irregularly shaped depressions spaced randomly or clustered on casting and are due to impurities collected at one or more vertices developed by the metal. Sand spots are caused by the metal washing particles from the runner system or mould walls, by excess turbulence in gating system, and by spurting of metal into the mould. These can be controlled by adopting proper moulding, gating and melting techniques.

6. Swell: It refers to the condition of enlargement of mould cavity when the molten metal is poured into the mould. It is caused, either by insufficient ramming or by pouring the metal too rapidly.

7. Shrinkage: It refers to the condition of voids in the casting resulting from concentrated contraction of the metal during solidification. It may be caused by improper location of gates and runners, poor design and inadequate filleting of corners.

8. Hot-Tears: There are the cracks having ragged edges due to tensile stresses during solidification. It is due to the discontinuity in the metal casting resulting from hindered contraction, occurring just after the metal has solidified.

It is caused by excessive mould hardness of ramming, high dry and hot strength, improper metallurgical and pouring temperature controls, and provision of insufficient fillets or brackets at the junctions of sections.

9. Cold-Cracks: These are similar to hot tears except that discontinuity is less and defect occurs below 270 °C.

10. Cold-Shots or Surface Laps: These are external defects caused by two streams of metals that are too cold to fuse properly; these can occur due to slow pouring, poor design and small gate; and can be controlled by the use of hotter metal using streamlined spines to give smoother flow. In this defect, small shot-like spheres of metal are almost distinct from casting.

11. Lifts and Shifts: They are external defects in castings caused due to misalignment of pattern parts, flask equipment, poor fitting of mould jackets and improper handling of moulds.

12. Sponginess or Honey-Combing: It is also an external defect, consisting of a number of cavities in close proximity. It is caused by dirt or swarm held in the molten metal, imperfect skimming and poor quality of molten metal.

13. Displaced Cores: These occur due to the buoyancy of cores in molten metal. Cores should be firmly anchored. In long cores, bending can be taken care of by using stiff core irons, and chaplets placed correctly.

14. Misplaced Cores: These result in unequal thickness of casting and occur due to moulder not checking up the various thicknesses, when finally assembling the mould and cores.

15. Pour-Short: It refers to the condition of incomplete filling of mould due to insufficient metal in the ladle and interruptions during pouring operation.

16. Gas Porosity: These are the rounded voids with smooth walls and occur due to gases dissolved in metal during melting and pouring. Imperfect feeding causes angular voids with dendrite arms protruding into the voids. Fine micro-porosity is observed in non-ferrous metals and occurs due to gas content and metal shrinkage.

17. Run-Outs: Drainage of metal from the cavity is called run-out. It gives incomplete casting and is caused by too large pattern, uneven match plate surfaces, inadequate mould weights and clamps, and excessive pouring pressure.

18. Metal Penetration: It refers to the condition of penetration of metal in the interstices of the sand grains. It causes a fused aggregate of metal and sand on the surface of casting which results in rough surface finish. It is caused by soft ramming, too coarse mould and core sand, and excessive metal temperature.

19. Fins: A thin projection of metal not intended as a part of casting is called fin. Fins usually occur at the parting of mould and core section. These are caused by run out of

metal, poor fittings of moulds and cores, high metal pressure, and insufficient weights and clamps.

20. Internal Air Pockets: These are caused by pouring boiling metal or rapid pouring of molten metal in the mould.

21. Dross or Sand Inclusion: These are oxides of other reaction products of metal being cast and these should be removed from the ladle before pouring metal. These defects are caused by improper control of melting and pouring, gating design and moulding sand practice.

Slag or dross inclusions can be prevented from entering from the ladle, by skimming before pouring or using bottom-pouring ladles, employing pouring basins so that any slag or dross entering from the ladle will rise and not pass into the runner system, designing the runner system to exert additional skimming action on the flowing metal, streamlining the runner system will minimise any tendency to entrap air or form dross or slag inclusions during pouring.

Careful control of mould permeability and gas content reduce danger of entrapping air or mould gases and minimize any tendency to form dross from mould-metal reactions.

22. Misruns: Misruns may be present in the form of improperly filled corners and mould cavities. These occur because of low pouring temperature, lack of fluidity of the metal, too small gates, too many restrictions in gating system etc. Another defect called the cold shot occurs when two cold streams of molten metal meet at the junction of a mould cavity and do not fuse together and thus the mould is not properly filled with metal.

23. Pinholes and Gas Holes: Pin holes are numerous, very small holes visible on the surface of casting after it has been cleaned by shot-blasting. Pinholes are caused by high- moisture and gas-producing material in sand due to faulty metal. Gas holes appear when the metal is machined or cut into sections. These also occur due to moisture and faulty metal.

24. Seams: These refer to defects at the junctions of two streams of metal.

25. Distortion: It occurs due to contraction stresses.

26. Drawing: It results in minute or fairly large holes with a black surface. It is associated with the contraction of metal in the mould. It is common with thick bosses which remain liquid after the surrounding metal has solidified. Such portions should be surrounded by chills to promote rapid cooling.

27. Some other minor casting defects are drops, crushes, cuts and washes. Drop occurs when the upper surface of mould cracks and pieces of sand fall into the molten metal.

Drop may occur due to low green strength, low mould hardness, using hot sand, insufficient reinforcement.

Defects Resulting from Incomplete Feeding

Solidification shrinkage is the biggest cause for many of casting defects.

External Imperfections

These appear in the form of localized cavities at unfed hot spots in the casting (depressed regions on cope and upper surfaces).

Whenever feeding is grossly inadequate, internal unsoundness is usually indicated by some external imperfection, wall punctures, deformation by dishing at the weakest point, elongated wormholes appearing at the riser on cope surfaces, defects resembling collections of dross where cope surfaces are wrinkled and drawn inward, and small voids in the form of pin holes.

Internal Imperfections

Like gross shrinkage, centre-line shrinkage, micro-porosity results from solidification shrinkage:

- Centre-Line Shrinkage: Centre-line shrinkage is a narrow, more or less continuous void sometimes found along the centre line of castings with extensive plate like sections.

- This defect is found only in alloys like steel which freeze over a relatively narrow temperature range.

- Interdendritic Shrinkage (Micro Porosity): Alloys that freeze over a wider temperature interval tend to exhibit this defect when improperly fed and also due to dissolved gases.

- Internal Hot Tearing: Internal hot tearing occurs due to improper feeding. 'Internal hot tears' are radially disposed discontinuities inside castings, emanating from low density area. These are disclosed by radiography. The discontinuities resemble external hot tears, except that they are radial rather than roughly parallel.

Advantages and Limitations

There are various types of casting processes, each with its own set of benefits and disadvantages. Below, you will find the different casting processes, each with its advantages, disadvantages and recommended application.

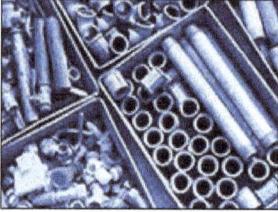

Investment Casting

Also known as lost wax casting, investment casting is a process commonly applied in cases where both solid parts and complex, hollow cores are required.

Advantages

- It is able to deliver close dimensional tolerances.

- Both ferrous and non-ferrous metals can be casted using investment casting.

- It delivers a good as-cast finish.

- With investment casting, complex shapes, intricate core sections, finer details and thinner walls are possible.

- It offers a flexibility in design and is a useful process for casting alloys that are difficult to machine.

Disadvantages

- Investment casting has a higher cost associated with it. In fact, it is more expensive than Sand Casting, Permanent Mold and the Plaster Casting process.

- When compared to other types of casting processes, investment casting needs a longer product-cycle time.

- There is a limitation on the size of parts that can be casted.

The advantages of this casting process overshadow costs when permanent mold and sand casting cannot deliver the desired complexity needed. Costs are also reduced by the quality surface finish investment casting delivers, reducing the cost of both machining and tooling.

Sand Casting

A process typically relying on silica-based materials, sand castings process involves finely ground, spherical grains tightly packed together into a smooth molding surface.

Advantages

- This is the least expensive process when producing small quantities (normally less than 100) and also boasts with the least expensive tooling.

- With sand casting, manufacturers are able to cast very large parts.

- Both ferrous and non-ferrous metals can be casted using this process.

- A low post-casting tooling cost.

Disadvantages

- Sand casting's dimensional accuracy is less than that delivered by other processes.

- This process requires large tolerances.

- The surface finish for ferrous casts delivered by this process usually exceeds 125 RMS.

- Castings produced with this proceed usually exceeds the calculated weight.

The advantages of sand casting process are more beneficial in cases where strength to weight ratios allows for it. Yielding a lower degree of accuracy, it does on the other hand offer low machining costs.

Die Casting

Die casting involves the molding of materials under high pressure, and consists of Cold-Chamber Conventional Die Casting, Hot-Chamber Conventional Die Casting and Multi-Slide Hot-Chamber Die Casting for aluminum, brass, magnesium and zinc.

Advantages

- Die casting is able to deliver parts that have a good dimensional tolerance.

- Parts produced using die casting require a minimal amount of post machining.

- This casting process also delivers an excellent part to part consistency, ideal for large production scales.

- It is a cost effective process when used for a high volume production run.

- This process is suitable for metals with a relatively low melting point, such as aluminum, lead, magnesium, zinc and some copper alloys.

Disadvantages

- This process is only an economically sound option for a large production quantity, as the tooling costs for die casting are expensive.

- In this process, it is difficult to guarantee minimum mechanical properties and is thus do not function as structural parts.

- This process is also not recommended for hydrostatic pressure applications.

- There is a limit on the size of parts that can be casted. It is a suitable process for castings of about up to 75 pounds.

While it is an economically viable option for large production runs, die cast parts' mechanical properties cannot be assured, resulting in these parts not having a structural function. It is however a good process to opt for in cases where a large quantity of the part is needed, the parts produced will not have a structural function and in cases where parts will not be subjected to hydrostatic pressure.

Permanent Mold Casting

Permanent mold casting is a process used in which permanent molds consists of mold cavities, machined into metal die blocks and appropriate for repetitive use.

Advantages

- This type of casting process is less expensive than die casting or investment casting.

- Castings delivered using this process are dense and pressure tight.

- Permanent mold casting is able to deliver a closer dimensional tolerance than sand casting is able to.

- The repeated uses of molds are possible.

- It has a rapid production rate combined with a low scrap rate.

Disadvantages

- This process is only able to cast non-ferrous metals.

- Permanent mold casting has a higher cost of tooling than sand casting has.

- It becomes less competitive with sand casting in cases where three or more sand cores are required in the process.

- Because of the high tooling cost associated with permanent mold casting, it is only a financially viable option for high production runs.

- Limited to small castings with a simple design exterior.

These types of casting processes are used in cases where parts are subjected to hydrostatic pressure and are perfect when casting parts that have no cores, a low profile and is part of a large production run (usually more than 300).

Recent developments do allow for more complex castings, such the aluminum engine blocks and heads produced.

Plaster Casting

Plaster casting is a casting process with similarities to sand casting. Instead of using sand in the process, a mixture of water, gypsum and strengthening compounds are used.

Advantages

- Plaster casting is capable of delivering a closer dimensional tolerance than sand casting is able.

- It delivers a smooth, as-cast finish.

- Casting larger parts using this process is less expensive that it would be when using investment casting processes.

- Intricate shapes with finer details are possible.

- Thinner, as-cast walls are also delivered by this casting process.

Disadvantages

- Plaster casting requires a minimum of a 1 degree draft.

- It is a more expensive process when compared to permanent mold and sand casting.

- This process may require the frequent replacement of plaster molding materials.

Though a more expensive process than most sand casting processes, it is a more economically sound option when a good surface finish quality is needed.

References

- What-is-casting-working-process-advantages-disadvantages-terminology-and-application: mech4 study.com, Retrieved 1 August , 2019

- Types-of-casting-in-manufacturing-r4, manufacturing-technology, notes: mechanical-engg.com, Retrieved 9 May, 2019

- Casting-components: themetalcasting.com, Retrieved 8 August , 2019

- Different-types-of-pattern-allowance-in-casting: mech4study.com, Retrieved 31 March, 2019

- Allowances-commonly-provided-on-a-pattern-casting, metallurgy: yourarticlelibrary.com, Retrieved 14 July, 2019

- Metalcasting-operation: the library of manufacturing.com, Retrieved 17 May, 2019

- Types-of-casting-defects-metals-industries-metallurgy, casting, metallurgy: engineeringenotes.com, Retrieved 19 April, 2019

- Pros-and-cons-of-casting-processes: chinasavvy.com, Retrieved 5 February, 2019

Types of Casting

There are numerous types of casting. A few of these are metal casting, continuous casting, die casting, gravity casting, centrifugal casting, permanent mold casting, sand casting and investment casting. The diverse applications of these types of casting have been thoroughly discussed in this chapter.

Metal Casting

Metal casting is a process in which hot liquid metal is poured into a mold that contains a hollow cutout or cavity of the desired finished shape. The liquid metal is then left to solidify, which is removed from the mold, revealing the end product, or the "Casting Form".

Semi-permanent Mould Casting

It is similar to permanent mould casting except the difference that while permanent moulds use metallic cores, the semi-permanent casting employs sand cores. It is used where cored openings are so irregular in shape, with undercuts, recesses, etc. that solid metal cores would be difficult to withdraw from the solidified castings.

Both the types of moulds are made in two or more pieces, which when fitted and clamped together define the outline of the part to be cast as well as the gates and risers. Stationary or moveable cores are used to form holes of any desired shape in the casting. Mould thickness is usually 25- 50 mm.

Principal factors in the production of sound castings are:

- Mould must be designed with parting lines, gates, vents etc. so that molten metal can enter by gravity without turbulence. Vents should be so arranged, that the air in the mould is pushed ahead of the gradually rising level of molten metal.

- Gating thickness and external contours of the mould should be such as to make possible progressive solidification of the molten metal in unbroken sequence.

- Mould temperature must be controlled within a definite range.

- Minimum diameter of cored holes is 6 to 10 mm to permit required strength at elevated temperature.

- Inserts can be easily cast into casting.

- Minimum section thickness is around 4 mm for magnesium alloys, 3 mm for aluminium and copper alloys.

- Usual draft angles on external and internal surfaces are 3° and 2° respectively.

- Generous fillets should be provided in corners especially where heavy and thin sections meet.

Slush Casting

This method is a special application of permanent mould casting in which hollow castings are produced without the use of cores. Molten metal is poured into the metallic mould and allowed to solidify upto the required thickness. The mould is then turned over so that the remaining liquid metal falls out and castings of desired thickness can be obtained.

Normally small thickness castings of lead, zinc and low melting alloys are obtained by slush casting method. The thickness of casting depends upon the time for which the metal is allowed to solidify into the permanent mould.

Since control is not precise, this method is not adopted to other than ornamental parts and for parts where only external features of the casting are important and uniformity of thickness is not important. In order to facilitate the removal of casting, the moulds are made in two halves. Ornaments, statues, toys and other novelties are the examples of slush casting.

The casting by this method must be made of a relatively pure metal as most metals do not form a strong solid skin. Owing to obvious drawbacks, this method is used only on a limited scale for low melting point alloys like lead or zinc.

Pressed Casting

It is another method of producing hollow castings from permanent moulds but differs from gravity die casting and slush casting in operation. In this case a definite amount of molten metal is poured into the mould and then close fitting cores are pushed in the cavities so that the molten metal can be forced into the mould cavities.

When the metal sets into the cavities, the core is removed and hence we get a thin walled hollow casting. Pressed die castings are limited to ornamental articles. This method was developed by Carthias of France and hence it is also popular by the name of Carthias casting.

Squeeze Casting

It produces non-ferrous castings having mechanical properties comparable with forgings. After the liquid metal has been metered into the open half of the lower die, a closely fitting upper die moves down and compresses the liquid at a high pressure and this pressure is maintained during solidification. Complex shaped components can be produced by this process.

Centrifugal Casting

In this process the molten metal is poured in a mould and allowed to solidify while the mould is revolving, i.e. the metal solidifies under the pressure of centrifugal force. The centrifugal forces cause the metal to take up the impression of the mould cavity. The pressure is selective, that is, the greater force is exerted on the denser components.

This is of considerable benefit in eliminating non-metallic and gases during casting. This process is best suited for mass production. Cylindrical parts and pipes are most adaptable to this process. In centrifugal casting, the molten metal is subjected to centrifugal force due to which if flows in mould cavities easily and results in the production of high density castings.

On casting surfaces better details of numerals and numbers can be obtained and thin parts of high strength can be easily produced. The castings are produced with promoted directional solidification as the colder metal is thrown to outer side of the castings and the hotter metal nearer the axis of rotation which further acts as a feeder during solidification of metal.

No core is needed to form the hole in the middle. Centrifugal casting finds its best use in mass production operation. The use of the machinery and equipment for centrifugal casting can be justified only when a large quantity of identical castings are required.

Typical Liner Casting Machine

This is horizontal type of centrifugal machine. There are two rollers at bottom and two at the top. The mould is arranged between the rollers so as to revolve freely. At the end of the mould is fitted a gear which meshes with a gear on a motor driven shaft. Thus the mould can be driven at a constant, predetermined speed.

The ends of the hollow mould are partially closed by covers which can be easily detached when casting is to be pushed out of the mould. At both end covers a central hole is provided. From one side, the molten metal is poured from a ladle and from the other the hot gases escape out.

For casting pipes, it would be preferable to have inclined machine to facilitate the flow of metal from one end of mould to the other. In order to avoid the chilling effect, the moulds have to be generally preheated to the requisite temperature. After centrifugal casting, the castings are immediately subjected to proper heat treatment in order to have the desired qualities and properties in the casting.

Selection of Proper Speed in Centrifugal Casting

Selection of proper speed is very important factor as the centrifugal force exerted on the molten metal is directly proportional to the square of the speed. In centrifugal casting, it is a common practice to produce a centrifugal force about 60 to 75 times of gravity for sand moulds with horizontal axis, 40—60 times for metal moulds and about 100 times for moulds revolving about a vertical axis.

If centrifugal force is more, then longitudinal hot tears in the outer surface of the casting are produced, whereas with low centrifugal force, slipping or raining of the molten metal during casting is experienced.

Exact spinning speed is dependent on several factors like application and shape of casting. Thus it is very difficult to determine precisely the exact value of rotation.

Material Considerations

Grey iron is centrifugally cast in large tonnages because of its relatively low pouring temperature and fluidity. Centrifugal casting of steel has replaced the forging methods. The heavy non-ferrous alloys, especially the copper base alloys, are readily formed by centrifugal casting.

Light metals have been centrifugally cast to some degree, although in some cases, where the oxides are as dense as the metal, centrifugal force is of little or no value in eliminating the non-metallic. Centrifugal casting has proved to be an efficient and economical method for producing annular components in special composition and heavy walled tubing of the common alloys, particularly for those alloys which are very difficult to forge or roll.

The bonding between two metals by this process is said to be complete and continuous and thus it is best suited for producing parts having soft lining on hard metals or vice versa.

Advantages of Centrifugal Casting

- The castings produced are sounder with dense structure, cleaner and the foreign inclusions are eliminated completely (these being segregated at the inner surface). This calls for simplified inspection techniques.

- Mass production is possible with less rejections.

- Use of runners and risers and cores is eliminated.

- Mechanical and physical properties of castings are improved.

- Parts are produced closer to finished dimensions with consequent saving in machining.

- Thinner sections can be cast because of the pressure exerted on the metal.

- Any metal can be cast by this process.

Limitations of Centrifugal Casting

- The process is limited to only cylindrical and annular parts with a limited range of sizes.

- It involves high initial cost and requires skilled labour for its maintenance.

- Too high speed may result in surface cracks caused by high stresses set up in the mould.

Investment of Lost-wax Casting

This process is called the lost-wax process or precision casting. This process uses wax pattern which is subsequently melted from the mould, leaving a cavity having all the details of the original pattern. Castings obtained by this process have very close tolerances or the order of ± 0.005 mm. Generally this process is used for producing light and intricate parts. This process does not need a parting line or any form of split mould.

The process of investment casting consists to two stages. First of all a master pattern is made of steel or brass and it is replica of the part to be cast. Around it, a split mould is formed from gelatin or an alloy of low melting point.

This alloy is poured over the master pattern. After solidification master mould is obtained. This master mould is used for making the wax or lost-pattern.

The following are the materials used for preparing master mould:

- Plaster of Paris or gypsum products for non-ferrous castings.

- Ethyl silicate, sodium silicate and phosphoric acid for steel castings.

- Sometimes fine-grain silica sand is also used for preparing master mould.

The master mould is then filled with either liquid wax or thermoplastic polystyrene resin which when solidified forms a wax-pattern. This wax pattern is used for making the final casting. Then the process of investment of the pattern is followed which consists of casting the wax-pattern with slurry consisting of silica sand or graphite mixed with water.

Coarse sand is sprinkled over the wet slurry to form the quartz shell. This wax-pattern is used for making the final mould in the same fashion as the conventional moulding process. This mould is then dried in air for 2 to 3 hours and then baked in an oven so that the wax may melt out. When the temperature reaches 100 to 120 °C, the wax melts out and is collected through a hole in the bottom plate.

To improve the resistivity, the mould is further heated up to 1000 °C called the sintering of the mould and finally cooled to 100 °C for obtaining the castings. The castings can be obtained by gravity, pressure, vacuum or centrifugal operation.

After the metal is cooled, the plaster is broken away and gates and feeders are cut out. The castings so obtained are finally cleaned by sand blasting, grinding or other finishing processes.

The castings so obtained have good surface finish and are exact reproduction of master pattern. This process is used for making jewellery parts, dental castings, castings of satellite tools (which are difficult to be produced either by forging or machining), turbine blades, parts of motor cars and sewing machines, type writers, calculating machines and various other intricate parts.

Advantages of Lost-wax Casting

- High dimensional accuracy of the order of ± 0.08 mm can be attained.

- A very smooth surface of the casting (of the order of 0.015 to 0.025 mm r.m.s. value) without parting line, can be easily obtained.

- It is suitable for both high and low melting point alloys since the ceramic material can be selected to have the appropriate refractory properties and bonded with any desired agent to give the required strength and permeability.

- It is a flexible process and can reproduce surface details and dimensions with precision, especially for high melting point alloys.

- Undercuts and other shapes, which would not allow the withdrawal of a normal pattern are easily provided. No cores or loose pieces are required.

- Machining of intricate parts can be eliminated.

- Very thin sections of the order of 3/4 mm can be cast easily.

- Die casting can be replaced when short runs are involved.

- Castings are sound and have large grains as the rate of cooling is slow.

- It represents the only method suitable for manufacture of precision shaped castings of high melting point metals which would cause too rapid die failures in normal die casting process.

Limitations of Lost-wax Casting

- It is an expensive process and hence is adopted only where small number of intricate and highly accurate parts particularly high melting point alloys is to be manufactured.

- This process is suitable for small size parts.

- This presents some difficulty where cores are to be used.

Frozen-mercury Moulding (Mercast Process)

In this process frozen mercury is used for the production of precision castings. In this case, the metal mould is prepared to the necessary shape with gates and sprue-holes. It is then placed in cold bath and filled with acetone (which acts as a lubricant). Mercury is poured into it and freezing of mercury takes place at 20 °C after about 10 minutes of pouring.

The patterns are then removed and are given dipping's in a cold ceramic slurry bath, until a shell of about 3 mm is built up. Mercury is then melted and removed at room temperature. The shell is dried and heated at high temperature to form a hard permeable shape. The shell is then placed in a flask, surrounded by sand, preheated and filled with metal. After solidification of metal, the castings can be removed.

Both the ferrous and non-ferrous metals can be cast by this process, but its application is limited for commercial use due to high cost of casting process.

Castings obtained by this process have the following characteristics:

- Very accurate details can be obtained even in intricate shapes.

- The surface finish is excellent and machining or cleaning cost is minimum.

- The accuracy obtained by this process is of the order of 0.002 mm per mm.

- Both ferrous and non-ferrous metals can be cast (maximum pouring temperature being around 1650 °C). However the cost of castings is high.

Plaster Mould Casting

For casting silver, gold, aluminium, magnesium, copper and alloys of those metals (particularly brass and bronze), plaster of Paris or gypsum ($CaSo_4.nH_2o$) is extensively used. Gypsum is particularly used for investment casting or for cope and drag moulding. For preparing the mould a slurry is used which consists of 100 parts of metal casting plaster and 160 parts of water.

It is important to note that plaster is to be added to water and not water to plaster of Paris. They are then stirred slowly to creamy consistency. This slurry is poured over a carefully made match plate type pattern, usually of metal. The mould is vibrated slightly to ensure plaster's filling all small cavities. The initial setting takes place at room temperature after few minutes of pouring of slurry and then the pattern can be removed.

Sometimes, the initial setting time is decreased by heating or by adding a small quantity of terra-alba. Copes and drags are made simultaneously on separate lines and dried in ovens at 200—425 °C, until all free and combined moisture is removed. Normally 20 hours is the time for drying purposes.

In order to prevent the cracking of moulds, 20 to 30% talc is added to the plaster while mixing. In addition other compounds such as terra-alba or magnesium oxide are added to reduce the initial setting time.

Sometimes lime or cement is also added to control the expansion of plaster during caking. Mould sections obtained by this process are very fragile and require care in assembling.

Characteristics and Advantages of Plaster Mould Casting

- The dimensional accuracy obtained by this process is of the order of 0.008 to 0.01 mm per mm.

- Because of no sand or other inclusions, excellent surface finish, which neither requires machining nor grinding, can be obtained.

- Advantage of this process is that non-ferrous castings having intricate and thin sections can be obtained with good dimensional accuracy and excellent surface finish. Because of low thermal conductivity of plaster the metal does not chill rapidly and thus very thin sections can be cast.

Limitations of Plaster Mould Casting

- Its application is limited to non-ferrous castings as sulphur of gypsum reacts chemically with ferrous metal at high temperature, giving very bad casting surface.

- Since the metal moulds are used, the plaster casting possesses low permeability because the combined water and moisture cannot be fully taken out. The moulds are not permanent. They are destroyed when the castings are removed.

Antioch Casting

This process is a further application of plaster casting. It was first development by Monis Beam for making special engineering parts of complex shape requiring minute details and thin sections. For preparing the mould a creamy slurry is obtained by adding water to a dry mixture of gypsum, sand, asbestos, talc and sodium silicate.

This slurry is piped by hose into metal core-boxes or cope and drag flasks fitted to special match plates. After the initial setting which is somewhat faster than plaster moulding, the patterns are drawn and then the mould is assembled in a green condition. After keeping them at room temperature for nearly 6 hours, they are autoclaved in steam at about 0.7 kg/cm² pressure, cured in air and finally dried in an oven at temperature of 230 °C for 10 hours.

This autoclaving develops a special permeable structure in the mould but greatly reduced dry strength. Drying in oven removes the fee and combined water and hence good permeability is obtained.

Advantages of the Antioch Process

- The major advantage of this process is that it develops a high degree of permeability in plaster mould and hence making it easier to obtain fine-details by allowing any moisture and other gases present to escape.

- Good finish and dimensional accuracy can be obtained even in large castings.

- This process lends itself to incorporate chills in the moulds, which can be used to control the metallurgical properties.

Continuous Casting

Continuous casting has proved itself to be a most economical way of casting wherever feasible and several methods have been devised and successfully used. In this process the molten metal is continuously poured into a mould around which there are facilities for rapidly chilling the metal to the point of solidification. The solidified metal is then continuously removed from the mould at the calculated rate.

Chill Casting

It is used, where very hard outer surfaces and wear resistant castings are required. This process is nearly similar to sand casting. Moulds are made of sand or cast iron and for the purpose of chilling the cast iron.

Metallic chills are used at outer surfaces so that the rate of cooling increases and hence hardness increases. Where hardness is of extreme importance, metallic moulds are used as in the case of railway brake shoe. In order to reduce the excessive chilling effect, moulds are preheated.

In case of bushes and bearings, the inner surfaces of the holes should be hard and wear resistant and in order to fulfil the above requirement core chills are used in the moulds. The core chills help in increasing the cooling rate of the bored surfaces by coming in contact with the chills. Extensive chills are used to reduce the possibility of the defect called hot-tear.

The rate of cooling has a considerable effect upon the hardness of the surface; greater the rate of cooling the lesser amount of carbon will come out in graphite state. In other words, carbon will be in the combined form and hence casting will be hard.

The examples of chill castings are wheel tread, railway brake shoe, tram-car wheels, crusher jaw, chilled rolls used in rolling mills and sideways of machine tools.

Continuous Casting

Continuous casting, also referred to as strand casting, is a process used in manufacturing industry to cast a continuous length of metal. Molten metal is cast through a mold, the casting takes the two dimensional profile of the mold but its length is indeterminate. The casting will keep traveling downward, its length increasing with time. New molten metal is constantly supplied to the mold, at exactly the correct rate, to keep up with the solidifying casting. Industrial manufacture of continuous castings is a very precisely calculated operation. Continuous casting can produce long strands from aluminum and copper, also the process has been developed for the production of steel.

Process of Continous Casting

Molten metal, from some nearby source, is poured into a tundish. A tundish is a container that is located above the mold, it holds the liquid metal for the casting. This particular casting operation uses the force of gravity to fill the mold and to help move along the continuous metal casting. The tundish is where the operation begins and is thus located high above ground level, as much as eighty or ninety feet. As can be seen, the continuous casting operation may require a lot of space.

It is the job of the tundish to keep the mold filled to the right level throughout the manufacturing operation. Since the metal casting is constantly moving through the mold, the tundish must always be supplying the mold with more molten metal to compensate.

The supplying of metal to the mold is not only going on throughout the entire manufacturing operation, it must be carried out with accuracy. A control system is employed to assist with this task. Basically the system can sense what the level of molten metal is, knows what the level should be, and can control the pouring of the metal from the tundish to ensure the smooth flow of the casting process. Although the tundish can typically hold several thousand pounds of metal, it too must be constantly supplied from the source of molten material.

The tundish also serves as the place where slag and impurities are removed from the melt. The high melting point and reactive nature, at high temperatures, has always made steel a difficult material to cast. When a manufacturing operation is continuously casting steel, the reactivity of the molten steel to the environment needs to be controlled. For this purpose, the mold entrance may be filled with an inert gas such as argon. The inert gas will push away any other gases, such as oxygen, that may react with the metal. There is no need to worry about the inert gas reacting with a molten metal melt, since inert gases do not react with anything at all.

The metal casting moves quickly through the mold, in the continuous manufacture of the metal part. The casting does not have time to solidify completely in the mold. As can be remembered from our discussion on solidification, a metal casting will first solidify from the mold wall, or outside of the casting, then solidification will progress inward. The mold in the continuous casting process is water cooled, this helps speed up the solidification of the metal casting. As stated earlier, the continuous casting does not completely harden in the mold. It does, however, spend enough time in the water cooled mold to develop a protective solidified skin of an adequate thickness on the outside.

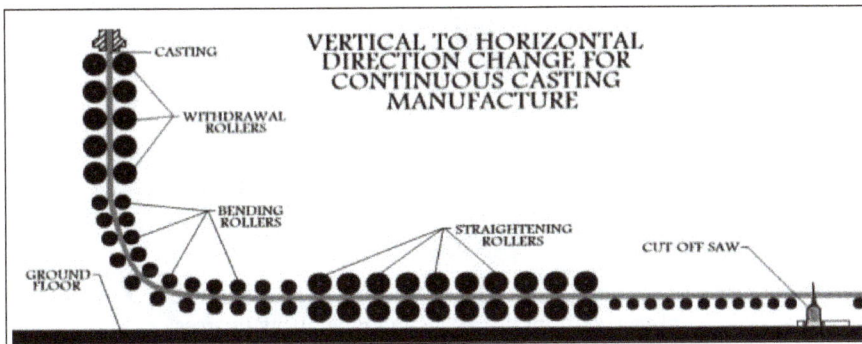

Vertical to horizontal direction change for continuous casting manufacture.

The long metal strand is moved along at a constant rate, by way of rollers. The rollers help guide the strand and assist in the smooth flow of the metal casting out of the mold and along its given path. A group of special rollers may be used to bend the strand to a 90 degree angle. Then another set will be used to straighten it, once it is at that angle. Commonly used in manufacturing industry, this process will change the direction of flow of the metal strand from vertical to horizontal.

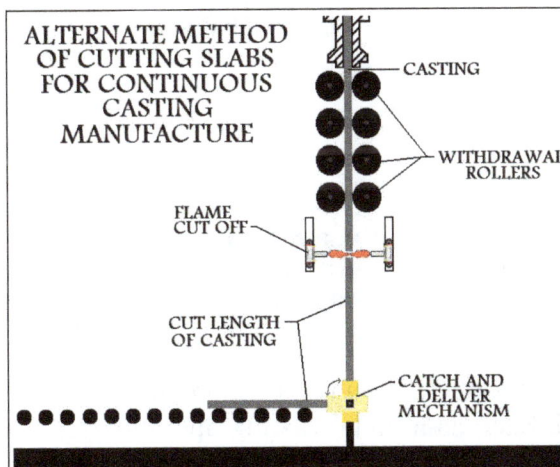

Alternate method of cutting slabs for continuous casting manufacture.

The continuous casting can now travel horizontally as far as necessary. The cutting device, in manufacturing industry, is typically a torch or a saw. Since the metal casting does not stop moving, the cutting device must move with the metal casting, at the same speed, as it does its cutting. There is another commonly used setup for cutting lengths of metal casting strand from a continuous casting operation. This particular manufacturing setup eliminates the need for bending and straightening rollers. It does, however, limit the length of metal casting strand that may be produced, based in a large part on the height of the casting floor where the mold is located.

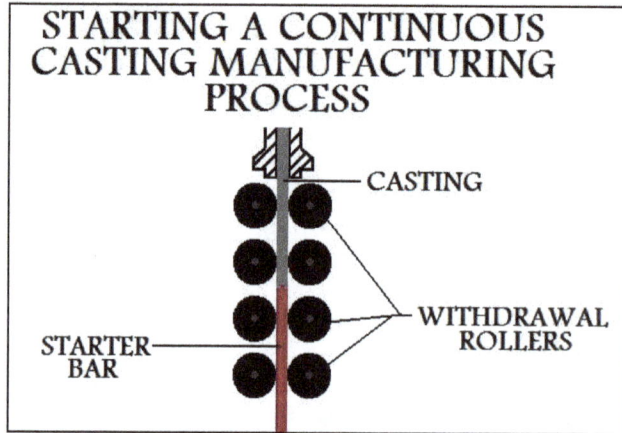

STARTING A CONTINUOUS CASTING MANUFACTURING PROCESS

CASTING

WITHDRAWAL ROLLERS

STARTER BAR

Starting a continuous casting manufacturing process.

There needs to be an initial setup for a continuous casting operation, since you can not just pour molten metal through an empty system to start off the process. To begin continuous casting manufacture, a starter bar is placed at the bottom of the mold. Molten material for the metal casting is poured into the mold and solidifies to the bar. The bar gives the rollers something to grab onto initially. The rollers pull the bar, which pulls along the continuous casting.

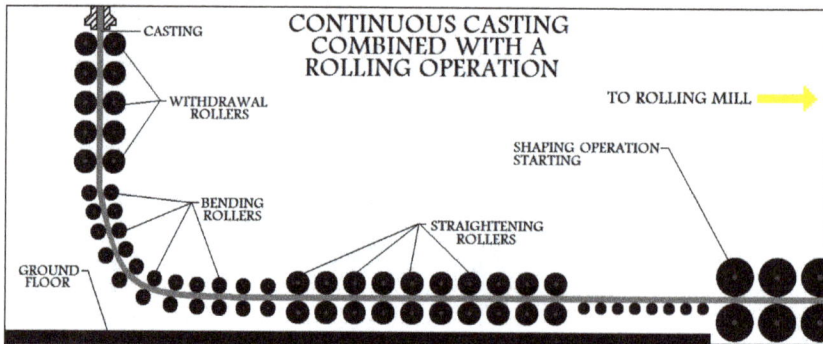

CASTING

CONTINUOUS CASTING COMBINED WITH A ROLLING OPERATION

TO ROLLING MILL ➡

WITHDRAWAL ROLLERS

SHAPING OPERATION STARTING

BENDING ROLLERS

STRAIGHTENING ROLLERS

GROUND FLOOR

Continuous casting combined with a rolling operation.

In the manufacture of a product, often two or more different kinds of operations may need to be performed. Such as a metal casting operation followed by a metal forming operation. In modern commercial industry, the continuous casting process can be integrated with metal rolling. Do not confuse the rolling operation with the rolls used to guide the casting. The rolling operation is a forming process and it will change the metal it processes. Rolling of the metal strand, is the second manufacturing process and it must be performed after the casting operation. Continuous casting is very convenient in that the rolling mill can be fed directly from the continuously cast metal casting strand. The metal strand can be rolled directly into a given cross sectional shape such as an I beam. The rate of the rolling operation is synchronized with the speed that the continuous metal casting is produced and thus the two operations are combined as one.

Properties and considerations of manufacturing by continuous casting:

- Continuous casting manufacture is different from other metal casting processes, particularly in the timing of the process. In other casting operations, the different steps to the process such as the ladling of metal, pouring, solidification, and casting removal all take place one at a time in a sequential order. In continuous casting manufacture, these steps are all occurring constantly and at the same time.

- This process is used in commercial manufacture as a replacement to the traditional process of casting ingots.

- Piping, a common problem in ingot manufacture, is eliminated with the continuous casting process.

- Structural and chemical variations in the metal of the casting, often present in ingots, have been eliminated. When manufacturing with the continuous metal casting process, the casting's material will possess uniform properties.

- When employing continuous metal casting manufacture, the castings will solidify at 10 times the rate that a casting solidifies during ingot production.

- With less loss of material, cost reduction, higher productivity rate, and superior quality of castings, continuous casting manufacture is often the choice over ingot production.

- A continuous casting manufacturing process will take considerable resources and planning to initiate, it will be employed in only very serious industrial operations.

Die Casting

Die casting is a manufacturing process that can produce geometrically complex metal parts through the use of reusable molds, called dies. The die casting process involves the use of a furnace, metal, die casting machine, and die. The metal, typically a non-ferrous alloy such as aluminum or zinc, is melted in the furnace and then injected into the dies in the die casting machine. There are two main types of die casting machines - hot chamber machines (used for alloys with low melting temperatures, such as zinc) and cold chamber machines (used for alloys with high melting temperatures, such as aluminum). The differences between these machines will be detailed in the sections on equipment and tooling. However, in both machines, after the molten metal is injected into the dies, it rapidly cools and solidifies into the final part, called the casting.

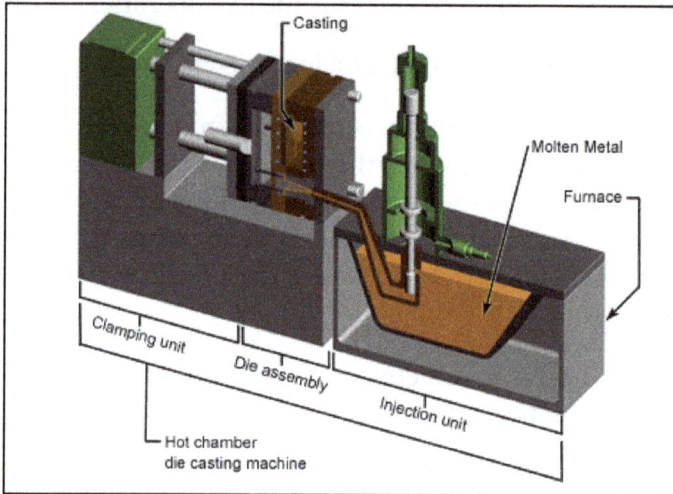

Die casting hot chamber machine overview.

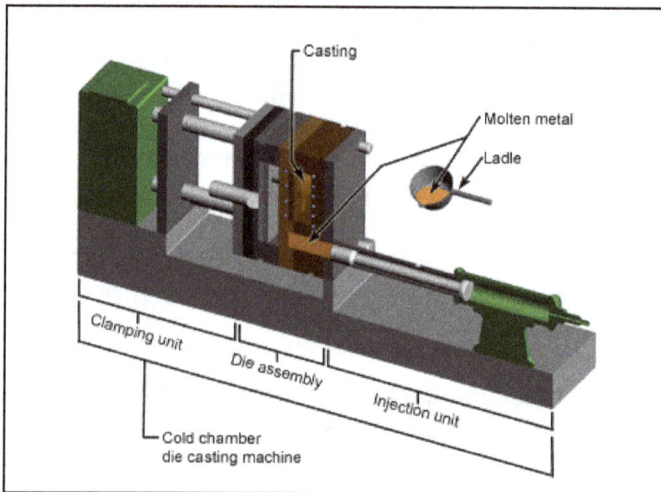

Die casting cold chamber machine overview.

The castings that are created in this process can vary greatly in size and weight, ranging from a couple ounces to 100 pounds. One common application of die cast parts are housings - thin-walled enclosures, often requiring many ribs and bosses on the interior. Metal housings for a variety of appliances and equipment are often die cast. Several automobile components are also manufactured using die casting, including pistons, cylinder heads, and engine blocks. Other common die cast parts include propellers, gears, bushings, pumps, and valves.

Process Cycle

The process cycle for die casting consists of five main stages, which are explained below. The total cycle time is very short, typically between 2 seconds and 1 minute.

- Clamping - The first step is the preparation and clamping of the two halves of the die. Each die half is first cleaned from the previous injection and then lubricated to facilitate the ejection of the next part. The lubrication time increases with part size, as well as the number of cavities and side-cores. Also, lubrication may not be required after each cycle, but after 2 or 3 cycles, depending upon the material. After lubrication, the two die halves, which are attached inside the die casting machine, are closed and securely clamped together. Sufficient force must be applied to the die to keep it securely closed while the metal is injected. The time required to close and clamp the die is dependent upon the machine - larger machines (those with greater clamping forces) will require more time. This time can be estimated from the dry cycle time of the machine.

- Injection - The molten metal, which is maintained at a set temperature in the furnace, is next transferred into a chamber where it can be injected into the die. The method of transferring the molten metal is dependent upon the type of die casting machine, whether a hot chamber or cold chamber machine is being used. Once transferred, the molten metal is injected at high pressures into the die. Typical injection pressure ranges from 1,000 to 20,000 psi. This pressure holds the molten metal in the dies during solidification. The amount of metal that is injected into the die is referred to as the shot. The injection time is the time required for the molten metal to fill all of the channels and cavities in the die. This time is very short, typically less than 0.1 seconds, in order to prevent early solidification of any one part of the metal. The proper injection time can be determined by the thermodynamic properties of the material, as well as the wall thickness of the casting. A greater wall thickness will require a longer injection time. In the case where a cold chamber die casting machine is being used, the injection time must also include the time to manually ladle the molten metal into the shot chamber.

- Cooling - The molten metal that is injected into the die will begin to cool and solidify once it enters the die cavity. When the entire cavity is filled and the molten metal solidifies, the final shape of the casting is formed. The die cannot be opened until the cooling time has elapsed and the casting is solidified. The cooling time can be estimated from several thermodynamic properties of the metal, the maximum wall thickness of the casting, and the complexity of the die. A greater wall thickness will require a longer cooling time. The geometric complexity of the die also requires a longer cooling time because the additional resistance to the flow of heat.

- Ejection - After the predetermined cooling time has passed, the die halves can be opened and an ejection mechanism can push the casting out of the die cavity. The time to open the die can be estimated from the dry cycle time of the machine and the ejection time is determined by the size of the casting's envelope and should include time for the casting to fall free of the die. The ejection mechanism must apply some force to eject the part because during cooling the part shrinks and adheres to the die. Once the casting is ejected, the die can be clamped shut for the next injection.

- Trimming - During cooling, the material in the channels of the die will solidify attached to the casting. This excess material, along with any flash that has occurred, must be trimmed from the casting either manually via cutting or sawing, or using a trimming press. The time required to trim the excess material can be estimated from the size of the casting's envelope. The scrap material that results from this trimming is either discarded or can be reused in the die casting process. Recycled material may need to be reconditioned to the proper chemical composition before it can be combined with non-recycled metal and reused in the die casting process.

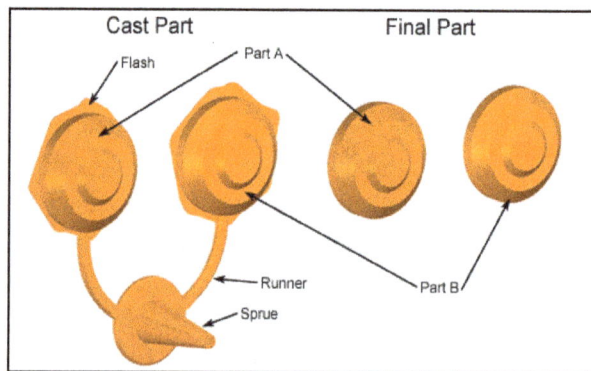

Die cast part.

Equipment

The two types of die casting machines are a hot chamber machine and cold chamber machine.

- Hot chamber die casting machine - Hot chamber machines are used for alloys with low melting temperatures, such as zinc, tin, and lead. The temperatures required to melt other alloys would damage the pump, which is in direct contact with the molten metal. The metal is contained in an open holding pot which is placed into a furnace, where it is melted to the necessary temperature. The molten metal then flows into a shot chamber through an inlet and a plunger, powered by hydraulic pressure, forces the molten metal through a gooseneck channel and into the die. Typical injection pressures for a hot chamber die casting machine are between 1000 and 5000 psi. After the molten metal has been injected into the die cavity, the plunger remains down, holding the pressure while the casting solidifies. After solidification, the hydraulic system retracts the plunger and the part can be ejected by the clamping unit. Prior to the injection of the molten metal, this unit closes and clamps the two halves of the die. When the die is attached to the die casting machine, each half is fixed to a large plate, called a platen. The front half of the die, called the cover die, is mounted to a stationary platen and aligns with the gooseneck channel. The rear half of the die, called the ejector die, is mounted to a movable platen, which slides along the tie bars. The hydraulically powered clamping unit actuates clamping bars that push this platen towards the cover die and exert enough pressure to keep it closed while the molten metal is injected. Following the solidification of the metal inside the die cavity, the clamping unit releases the die halves and simultaneously causes the ejection system to push the casting out of the open cavity. The die can then be closed for the next injection.

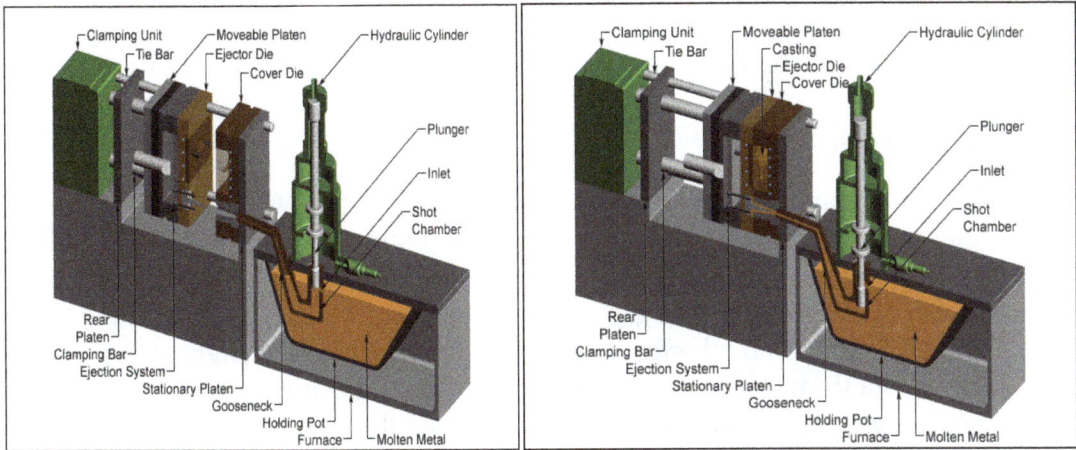

Hot chamber die casting machine – Opened.

Hot chamber die casting machine - Closed.

- Cold chamber die casting machine - Cold chamber machines are used for alloys with high melting temperatures that cannot be cast in hot chamber machines because they would damage the pumping system. Such alloys include aluminum, brass, and magnesium. The molten metal is still contained in an open holding pot which is placed into a furnace, where it is melted to the necessary temperature. However, this holding pot is kept separate from the die casting machine and the molten metal is ladled from the pot for each casting, rather than being pumped. The metal is poured from the ladle into the shot chamber through a pouring hole. The injection system in a cold chamber machine functions similarly to that of a hot chamber machine, however it is usually oriented horizontally and does not include a gooseneck channel. A plunger, powered by hydraulic pressure, forces the molten metal through the shot chamber and into the injection sleeve in the die. The typical injection pressures for a cold chamber die casting machine are between 2000 and 20000 psi. After the molten metal has been injected into the die cavity, the plunger remains forward, holding the pressure while the casting solidifies. After solidification, the hydraulic system retracts the plunger and the part can be ejected by the clamping unit. The clamping unit and mounting of the dies is identical to the hot chamber machine.

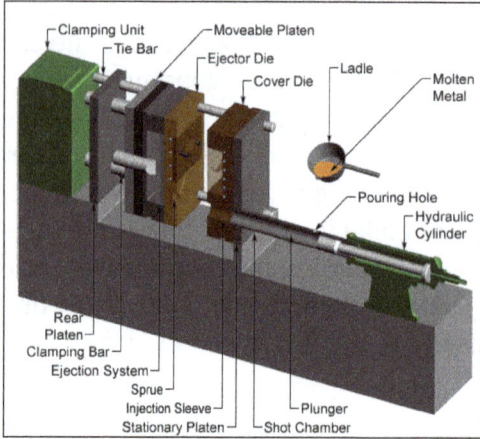

Cold chamber die casting machine – Opened.

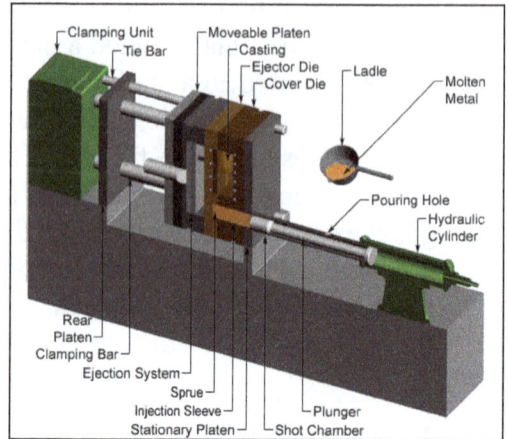

Cold chamber die casting machine – Closed.

Machine Specifications

Both hot chamber and cold chamber die casting machines are typically characterized by the tonnage of the clamp force they provide. The required clamp force is determined by the projected area of the parts in the die and the pressure with which the molten metal is injected. Therefore, a larger part will require a larger clamping force. Also, certain materials that require high injection pressures may require higher tonnage machines. The size of the part must also comply with other machine specifications, such as maximum shot volume, clamp stroke, minimum mold thickness, and platen size.

Die cast parts can vary greatly in size and therefore require these measures to cover a very large range. As a result, die casting machines are designed to each accommodate a small range of this larger spectrum of values. Sample specifications for several different hot chamber and cold chamber die casting machines are given below:

Type	Clamp force (ton)	Max. shot volume (oz.)	Clamp stroke (in.)	Min. mold thickness (in.)	Platen size (in.)
Hot chamber	100	74	11.8	5.9	25 × 24
Hot chamber	200	116	15.8	9.8	29 × 29
Hot chamber	400	254	21.7	11.8	38 × 38
Cold chamber	100	35	11.8	5.9	23 × 23
Cold chamber	400	166	21.7	11.8	38 × 38
Cold chamber	800	395	30.0	15.8	55 × 55
Cold chamber	1600	1058	39.4	19.7	74 × 79
Cold chamber	2000	1517	51.2	25.6	83 × 83

Tooling

The dies into which the molten metal is injected are the custom tooling used in this process. The dies are typically composed of two halves - the cover die, which is mounted onto a stationary platen, and the ejector die, which is mounted onto a movable platen. This design allows the die to open and close along its parting line. Once closed, the two die halves form an internal part cavity which is filled with the molten metal to form the casting. This cavity is formed by two inserts, the cavity insert and the core insert, which are inserted into the cover die and ejector die, respectively. The cover die allows the molten metal to flow from the injection system, through an opening, and into the part cavity. The ejector die includes a support plate and the ejector box, which is mounted onto the platen and inside contains the ejection system. When the clamping unit separates the die halves, the clamping bar pushes the ejector plate forward inside the ejector box which pushes the ejector pins into the molded part, ejecting it from the core insert. Multiple-cavity dies are sometimes used, in which the two die halves form several identical part cavities.

Die Channels

The flow of molten metal into the part cavity requires several channels that are integrated into the die and differs slightly for a hot chamber machine and a cold chamber machine. In a hot chamber machine, the molten metal enters the die through a piece called a sprue bushing (in the cover die) and flows around the sprue spreader (in the ejector die). The sprue refers to this primary channel of molten metal entering the die. In a cold chamber machine, the molten metal enters through an injection sleeve. After entering the die, in either type of machine, the molten metal flows through a series of runners and enters the part cavities through gates, which direct the flow. Often, the cavities will contain extra space called overflow wells, which provide an additional source of molten metal during solidification. When the casting cools, the molten metal will shrink and additional material is needed. Lastly, small channels are included that run from the cavity to the exterior of the die. These channels act as venting holes to allow air to escape the die cavity. The molten metal that flows through all of these channels will solidify attached to the casting and must be separated from the part after it is ejected. One type of channel that does not fill with material is a cooling channel. These channels allow water or oil to flow through the die, adjacent to the cavity, and remove heat from the die.

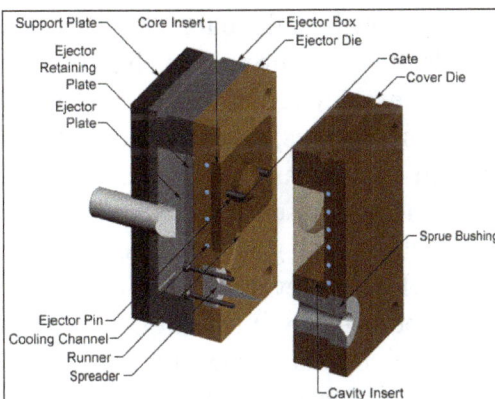

Die assembly – Open (Hot chamber).

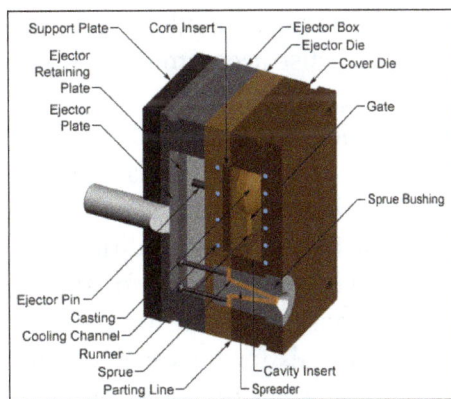

Die assembly – Closed (Hot chamber).

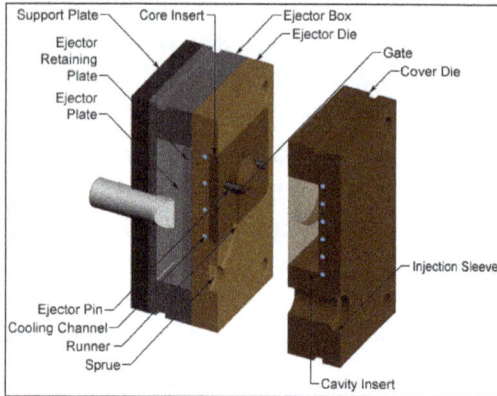

Die assembly – Opened (Cold chamber).

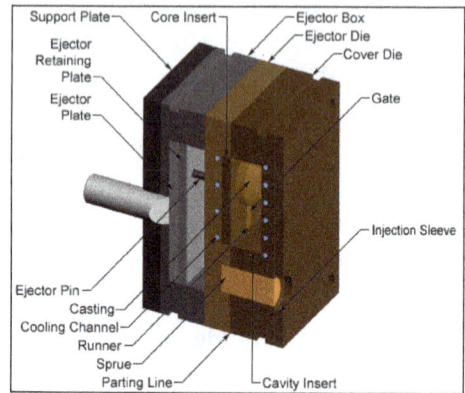

Die assembly – Closed (Cold chamber).

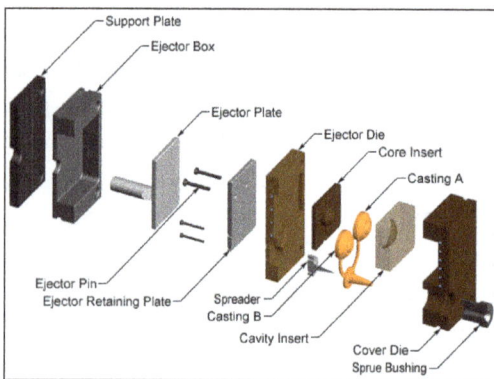

Die assembly - Exploded view (Hot chamber).

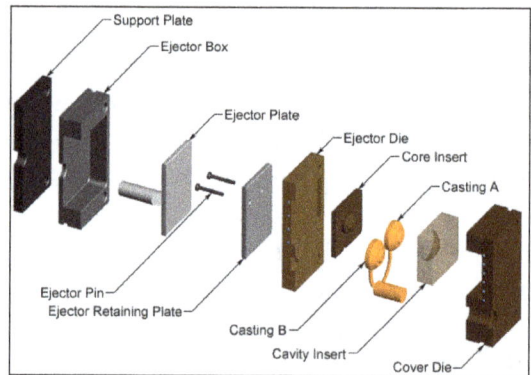

Die assembly - Exploded view (Cold chamber).

Die Design

In addition to these many types of channels, there are other design issues that must be considered in the design of the dies. Firstly, the die must allow the molten metal to flow easily into all of the cavities. Equally important is the removal of the solidified casting from the die, so a draft angle must be applied to the walls of the part cavity. The design of the die must also accommodate any complex features on the part, such as undercuts, which will require additional die pieces. Most of these devices slide into the part cavity through the side of the die, and are therefore known as slides, or side-actions. The most common type of side-action is a side-core which enables an external undercut to be molded. Another important aspect of designing the dies is selecting the material. Dies can be fabricated out of many different types of metals. High grade tool steel is the most common and is typically used for 100-150,000 cycles. However, steels with low carbon content are more resistant to cracking and can be used for 1,000,000 cycles. Other common materials for dies include chromium, molybdenum, nickel alloys, tungsten, and vanadium. Any side-cores that are used in the dies can also be made out of these materials.

Materials

Die casting typically makes use of non-ferrous alloys. The four most common alloys that are die cast are shown below, along with brief descriptions of their properties:

Materials	Properties
Aluminum alloys	• Low density. • Good corrosion resistance. • High thermal and electrical conductivity. • High dimensional stability. • Relatively easy to cast. • Requires use of a cold chamber machine.
Copper alloys	• High strength and toughness. • High corrosion and wear resistance. • High dimensional stability. • Highest cost. • Low die life due to high melting temperature. • Requires use of a cold chamber machine.
Magnesium alloys	• Very low density. • High strength-to-weight ratio. • Excellent machinability after casting. • Use of both hot and cold chamber machines.
Zinc alloys	• High density. • High ductility. • Good impact strength. • Excellent surface smoothness allowing for painting or plating. • Requires such coating due to susceptibility to corrosion. • Easiest to cast. • Can form very thin walls. • Long die life due to low melting point. • Use of a hot chamber machine.

The selection of a material for die casting is based upon several factors including the density, melting point, strength, corrosion resistance, and cost. The material may also affect the part design. For example, the use of zinc, which is a highly ductile metal, can allow for thinner walls and a better surface finish than many other alloys. The material not only determines the properties of the final casting, but also impacts the machine and tooling. Materials with low melting temperatures, such as zinc alloys, can be die cast in a hot chamber machine. However, materials with a higher melting temperature, such as aluminum and copper alloys, require the use of cold chamber machine. The melting temperature also affects the tooling, as a higher temperature will have a greater adverse effect on the life of the dies.

Possible Defects Causes
Flash

- Injection pressure too high.

- Clamp force too low.

Unfilled Sections

- Insufficient shot volume.

- Slow injection.

- Low pouring temperature.

Bubbles

- Injection temperature too high.

- Non-uniform cooling rate.

Hot Tearing

- Non-uniform cooling rate.

Ejector Marks

- Cooling time too short.

- Ejection force too high.

Many of the above defects are caused by a non-uniform cooling rate. A variation in the cooling rate can be caused by non-uniform wall thickness or non-uniform die temperature.

Design Rules
Maximum Wall Thickness

- Decrease the maximum wall thickness of a part to shorten the cycle time (injection time and cooling time specifically) and reduce the part volume.

Incorrect   Correct

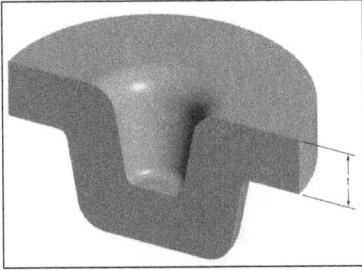 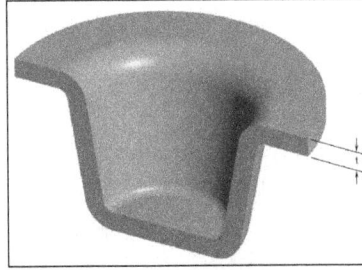

Part with thick walls.   Part redesigned with thin walls.

- Uniform wall thickness will ensure uniform cooling and reduce defects.

Incorrect   Correct

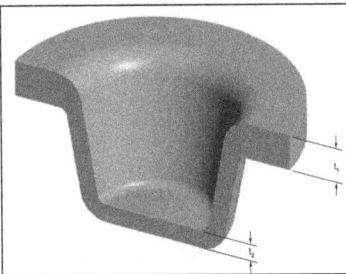 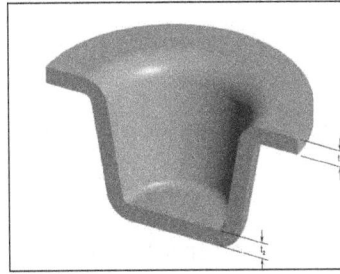

Non-uniform wall thickness ($t_1 \neq t_2$).   Uniform wall thickness ($t_1 = t_2$).

Corners

- Round corners to reduce stress concentrations and fracture.

- Inner radius should be at least the thickness of the walls.

Incorrect   Correct

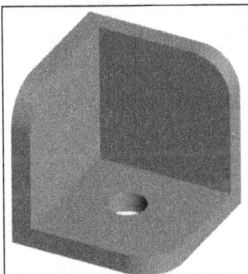 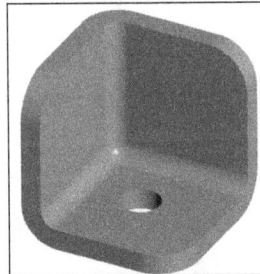

Sharp corner.   Rounded corner.

Draft

- Apply a draft angle to all walls parallel to the parting direction to facilitate re-moving the part from the die.

- ◦ Aluminum: 1° for walls, 2° for inside cores.

- ◦ Magnesium: 0.75° for walls, 1.5° for inside cores.

- ◦ Zinc: 0.5° for walls, 1° for inside cores.

Incorrect   Correct

   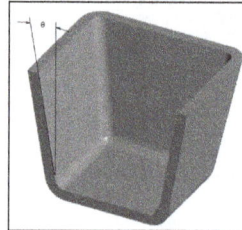

No draft angle.   Draft angle (q).

Undercuts

- Minimize the number of external undercuts.

  - ◦ External undercuts require side-cores which add to the tooling cost.

- Some simple external undercuts can be cast by relocating the parting line.

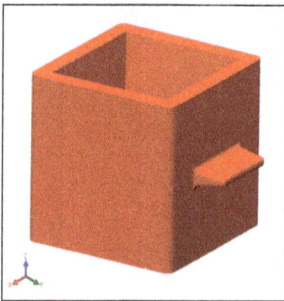   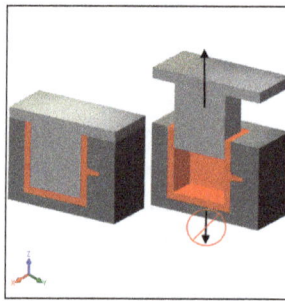   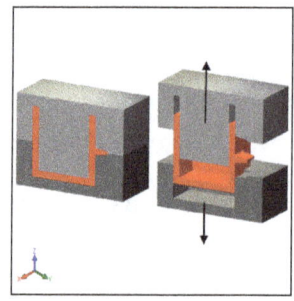

Simple external undercut.   Die cannot separate.   New parting line allows undercut.

- Redesigning a feature can remove an external undercut.

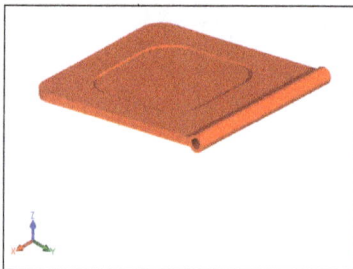   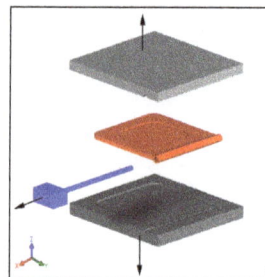

Part with hinge.   Hinge requires side-core.

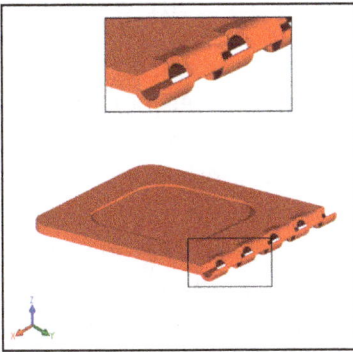

Redesigned hinge.

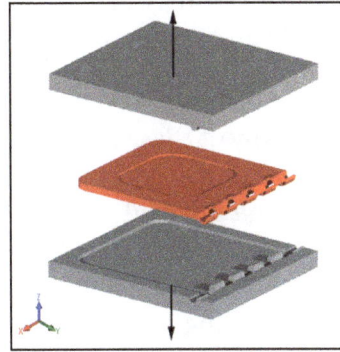

New hinge can be cast.

- Minimize number of side-action directions.

  ○ Additional side-action directions will limit the number of possible cavities in the die.

Cost Drivers

Material Cost

The material cost is determined by the weight of material that is required and the unit price of that material. The weight of material is clearly a result of the part volume and material density; however, the part's maximum wall thickness can also play a role. The weight of material that is required includes the material that fills the channels of the die. A part with thinner walls will require a larger system of channels to ensure that the entire part fills quickly and evenly, and therefore will increase the amount of required material. However, this additional material is typically less than the amount of material saved from the reduction in part volume, a result of thinner walls. Therefore, despite the larger channels, using thinner walls will typically lower the material cost.

Production Cost

The production cost is primarily calculated from the hourly rate and the cycle time. The hourly rate is proportional to the size of the die casting machine being used, so it is important to understand how the part design affects machine selection. Die casting machines are typically referred to by the tonnage of the clamping force they provide. The required clamping force is determined by the projected area of the part and the pressure with which the molten metal is injected. Therefore, a larger part will require a larger clamping force, and hence a more expensive machine. Also, certain materials that require high injection pressures may require higher tonnage machines. The size of the part must also comply with other machine specifications, such as clamp stroke, platen size, and shot capacity. In addition to the size of the machine, the type of machine (hot chamber vs. cold chamber) will also affect the cost. The use of materials with high melting temperatures, such as aluminum, will require cold chamber machines which are typically more expensive.

The cycle time can be broken down into the injection time, cooling time, and resetting time. By reducing any of these times, the production cost will be lowered. The injection time can be decreased by reducing the maximum wall thickness of the part. Also, certain materials can be injected faster than others, but the injection times are so short that the cost saving are negligible. Substantial time can be saved by using a hot chamber machine because in cold chamber machines the molten metal must be ladled into the machine. This ladling time is dependent upon the shot weight. The cooling time is also decreased for lower wall thicknesses, as they require less time to cool all the way through. Several thermodynamic properties of the material also affect the cooling time. Lastly, the resetting time depends on the machine size and the part size. A larger part will require larger motions from the machine to open, close, and eject the part, and a larger machine requires more time to perform these operations. Also, the use of any side-cores will slow this process.

Tooling Cost

The tooling cost has two main components - the die set and the machining of the cavities. The cost of the die set is primarily controlled by the size of the part's envelope. A larger part requires a larger, more expensive, die set. The cost of machining the cavities is affected by nearly every aspect of the part's geometry. The primary cost driver is the size of the cavity that must be machined, measured by the projected area of the cavity (equal to the projected area of the part and projected holes) and its depth. Any other elements that will require additional machining time will add to the cost, including the feature count, parting surface, side-cores, tolerance, and surface roughness.

The quantity of parts and material used will affect the tooling life and therefore impact the cost. Materials with high casting temperatures, such as copper, will cause a short tooling life. Zinc, which can be cast at lower temperatures, allows for a much longer tooling life. This effect becomes more cost prohibitive with higher production quantities.

One final consideration is the number of side-action directions, which can indirectly affect the cost. The additional cost for side-cores is determined by how many are used. However, the number of directions can restrict the number of cavities that can be included in the die. For example, the die for a part which requires 3 side-core directions can only contain 2 cavities. There is no direct cost added, but it is possible that the use of more cavities could provide further savings.

Types of Die Casting Machines

Hot-chamber Die Casting Machines

A hot-chamber die casting machine is shown in figure. The main Components of the machine includes a steel pot filled with the molten metal to be cast and a pumping system that consists of a pressure cylinder, a plunger, a gooseneck passage, and a nozzle.

With the plunger in the up position, the molten metal flows by gravity through the intake ports into the submerged hot chamber. When the plunger is pushed downward by the power cylinder, it shuts off the intake port.

Then, with further downward movement of plunger, the molten metal is forced through the gooseneck passage and the nozzle into the die cavity as shown in figure. The pressure of molten metal coming out from the nozzle is about 50 to 150 atmospheres per square inch.

The pressure is maintained after the cavity is full of molten metal, for a specific time to solidify the casting completely. Next, the two halves of the die are separated and before the cycle are repeated.

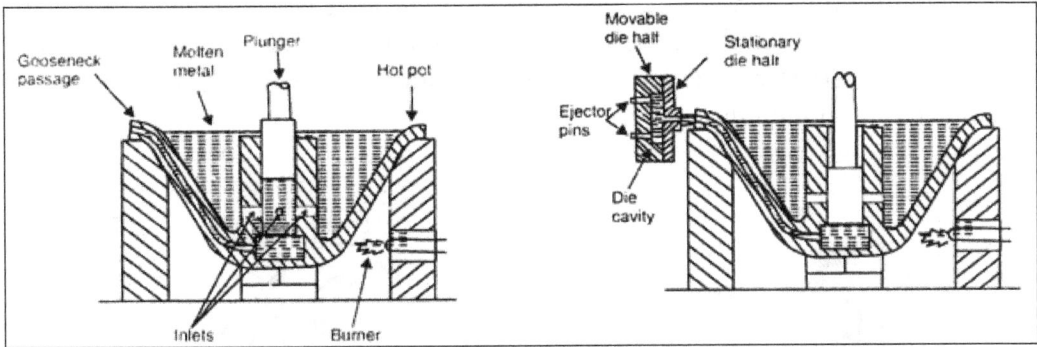

The hot-chamber die-casting machines: a) Filling the chamber. b) Metal forced into die-casting.

Advantages of Hot-chamber Die-casting

The advantages of hot-chamber die casting are numerous, some important to write here are:

- High production rates, especially when multi-cavity dies are used.

- Improved productivity and surface finish.

- Very close dimensional tolerances.

- Ability to produce intricate shapes with thin walls.

Limitations of Hot-chamber Die-casting

Nevertheless, the hot-chamber die casting has some limitations, these are:

- Only low-melting-point alloys (such as zinc, tin, lead, aluminum and like) can be cast because the components of the pumping system are in direct contact with the molten metal throughout the process.

- Also, it is usually only suitable for producing small castings that weigh less than 4.5 kg.

Cold-chamber Die Casting Machines

A Cold-chamber die casting machine is shown in figure. the molten metal is first ladled through the pouring hole of the shot chamber. The two halves of the die are closed and locked together. Next, the plunger moves forward to close off the pouring hole and forces the molten metal into the die cavity.

The pressure in shot chamber may go over 2000 atmospheric per square inch. After the castings have solidified, the two halves of the die are separated, and the casting, together with the gate and slag of excess metal, are ejected from the die, by means of ejector pins.

In cold-chamber die casting machines, the molten-metal reservoir is separate from casting machines, unlike the hot-chamber die casting machines. One shot of molten metal is ladled every stroke. The steel chamber (shot chamber) is too little to have any reaction with hot molten metal to be casted.

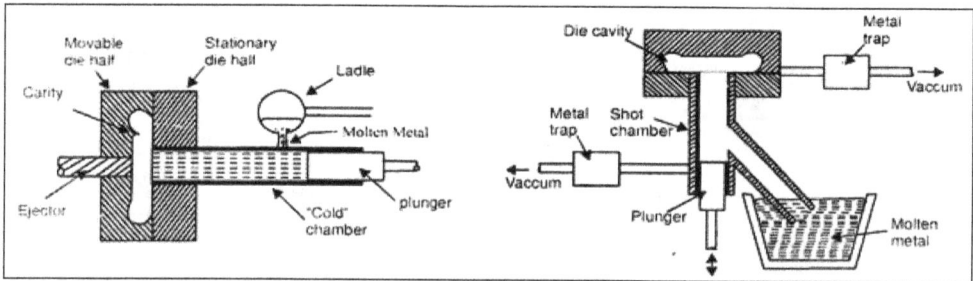

The cold-chamber die-casting machines: a)Horizontal cold-chamber. b)Vertical cold-chamber.

Advantages of Cold-chamber Die-casting

- Large parts weighing 20 kg can be produced by cold-chamber die casting.
- The process is very successful for casting aluminum and alloys, copper and alloys, and high- temperature zinc-aluminum alloys.
- Intricate shapes are easily made.

Limitations of Cold-chamber Die-casting

- A longer cycle time when compared with hot-chamber die casting.
- An auxiliary system for pouring the molten metal is needed.

Owing to the above limitations, Vertical cold-chamber machines were developed. A typical vertical cold- chamber machine is shown in figure. It has a transfer tube that is submerged into the molten metal. It is fed into the shot chamber by connecting the die cavity to a vacuum tank by means of a special valve. Then the molten metal is forced into the die cavity when the plunger moves upward.

Hot Die Casting

Hot chamber die casting is one of the two main techniques in the manufacturing process of die casting.

Hot Chamber Process

A similar characteristic of either die casting process is the use of high pressure to force molten metal through a mold called a die. Many of the superior qualities of castings manufactured by die casting, (such as great surface detail), can be attributed to the use of pressure to ensure the flow of metal through the die. In hot chamber die casting manufacture, the supply of molten metal is attached to the die casting machine and is an integral part of the casting apparatus for this manufacturing operation.

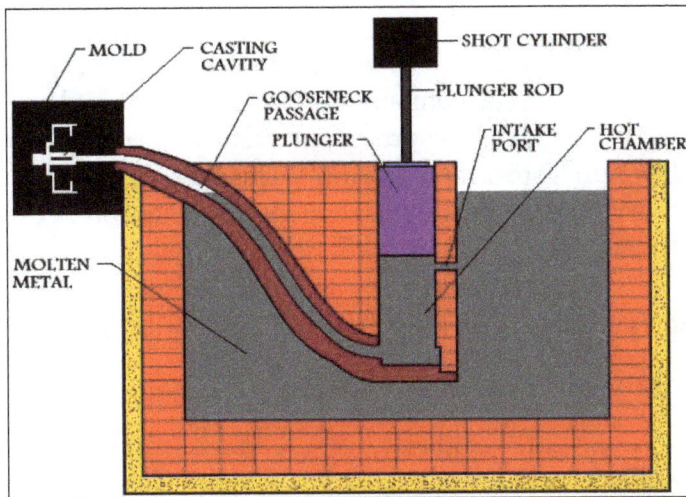

Hot chamber die casting.

The shot cylinder provides the power for the injection stroke. It is located above the supply of molten metal. The plunger rod goes from the shot cylinder down to the plunger, which is in contact with the molten material. At the start of a casting cycle, the plunger is at the top of a chamber (the hot-chamber). Intake ports allow this chamber to fill with liquid metal.

As the cycle begins, the power cylinder forces the plunger downward. The plunger travels past the ports, cutting off the flow of liquid metal to the hot chamber. Now there should be the correct amount of molten material in the chamber for the "shot" that will be used to fill the mold and produce the casting.

At this point the plunger travels further downward, forcing the molten metal into the die. The pressure exerted on the liquid metal to fill the die in hot chamber die casting manufacture usually varies from about 700psi to 5000psi (5MPa to 35 MPa). The pressure is held long enough for the casting to solidify.

In preparation for the next cycle of casting manufacture, the plunger travels back upward in the hot chamber exposing the intake ports again and allowing the chamber to refill with molten material.

Hot chamber die casting has the advantage of a very high rate of productivity. During industrial manufacture by this process one of the disadvantages is that the setup requires that critical parts of the mechanical apparatus, (such as the plunger), must be continuously submersed in molten material. Continuous submersion in a high enough temperature material will cause thermal related damage to these components rendering them inoperative. For this reason, usually only lower melting point alloys of lead, tin, and zinc are used to manufacture metal castings with the hot chamber die casting process.

Pressure Die Casting

Pressure die casting is a quick, reliable and cost-effective manufacturing process for production of high volume, metal components that are net-shaped have tight tolerances. Basically, the pressure die casting process consists of injecting under high pressure a molten metal alloy into a steel mold (or tool). This gets solidified rapidly (from milliseconds to a few seconds) to form a net shaped component. It is then automatically extracted.

Advantages of Pressure Die Casting

- Lower costs compared to other processes.

- Economical - typically production of any number of components from thousands to millions before requiring replacement is possible.

- Castings with close dimensional control and good surface finish.

- Castings with thin walls, and therefore are lighter in weight.

Types of Pressure Die Casting

- High Pressure Die Casting.

- Low Pressure Die Casting.

Depending upon the pressure used, there are two types of pressure die casting namely High Pressure Die Casting and Low Pressure Die Casting. While high pressure die casting has wider application encompassing nearly 50% of all light alloy casting production.

Currently low pressure die casting accounts for about 20% of the total production but its use is increasing. High pressure castings are must for castings requiring tight tolerance and detailed geometry. As the extra pressure is able to push the metal into more detailed features in the mold. Low pressure die casting is commonly used for larger and non-critical parts.

However, the machine and its dies are very costly, and for this reason pressure die casting is viable only for high-volume production.

High Pressure Die Casting

Here, the liquid metal is injected with high speed and high pressure into the metal mold. The basic equipment consists of two vertical platens. The bolsters are placed on these platens and this holds the die halves. Out of the two platens, one is fixed and the other movable.

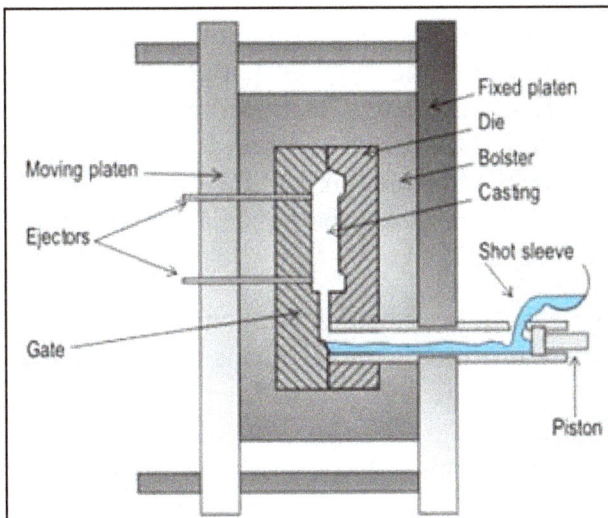

High Pressure Die Casting Process.

This helps the die to open and close. A specific amount of metal is poured into the shot sleeve and afterwards introduced into the mold cavity. This is done using a hydraulically-driven piston. After the metal has solidified, the die is opened and the casting eventually removed.

Types of High Pressure Die Casting

Both the processes are described below. The only difference between the two processes is the method being used to inject molten metal into the die.

Hot Chamber Process

The hot-chamber process is applicable only for zinc and other low melting point alloys that does not affect and erode metal pots cylinders and plungers. The basic components of a hot-chamber diecasting machine and die are illustrated below:

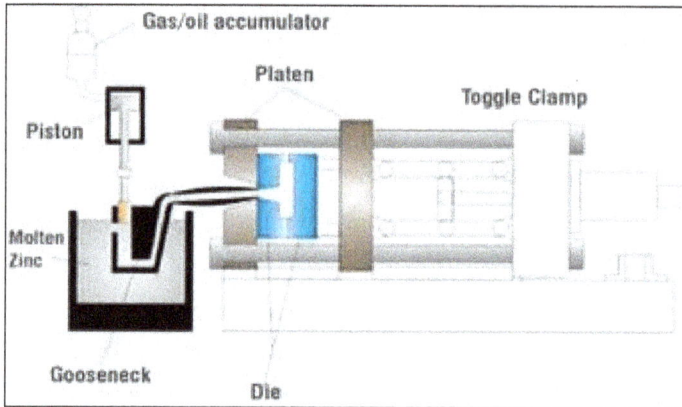

Hot Chamber Process.

The workings of a hot chamber process goes like this. The molten metal for casting is placed in the holding furnace at the required temperature adjacent to(sometimes as part of the machine itself) the machine. The injection mechanism is placed within the holding furnace and most of its part is in constant touch with the molten metal. When pressure is transmitted by the injection piston, the metal is forced through the gooseneck into the die. On the return stroke, the metal is drawn towards the gooseneck for the next shot.

This process ensures minimum contact between air and the metal to be injected. The tendency for entrainment of air in the metal during injection is also minimised.

Cold Chamber Process

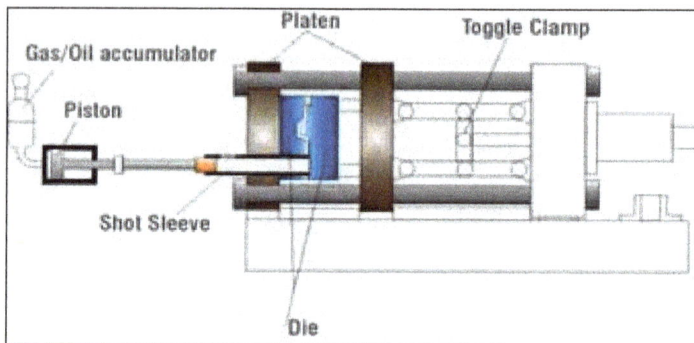

Cold Chamber Process.

The difference of this process with the hot-chamber process is that the injection system is not submerged in molten metal. On the contrary, metal gets transferred by ladle, manually or automatically, to the shot sleeve. The metal is pushed into the die by a hydraulically operated plunger. This process minimises the contact time between the injector components and the molten metal. Which extends the life of the components.

However the entrainment of air into the metal generally associated with high-speed injection can cause gas porosity in the castings. In the cold chamber machine, injection pressures over 10,000 psi or 70,000 KPa is obtainable. Generally steel castings along with aluminium and copper based alloys are produced by this method.

Low Pressure Die Casting

High quality castings, of aluminium alloys, along with magnesium and other low melting point alloys are usually produced through this process. Castings of aluminium in the weight range of 2-150 kg are a common feature.

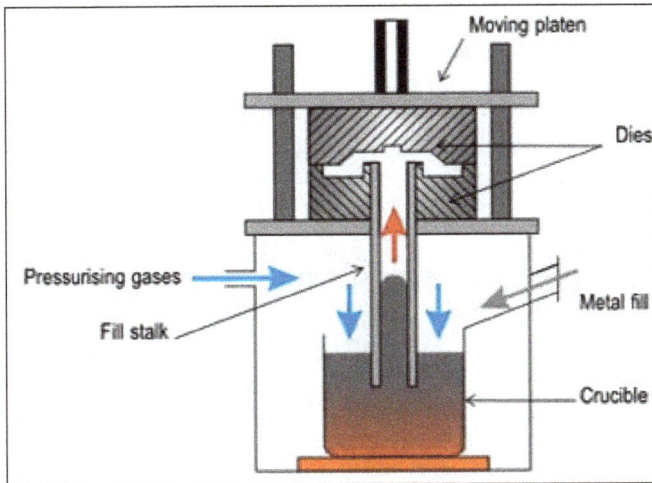

Low Pressure Die Casting Process.

The process works like this, first a metal die is positioned above a sealed furnace containing molten metal. A refractory-lined riser extends from the bottom of the die into the molten metal. Low pressure air (15 - 100 kPa, 2- 15 psi) is then introduced into the furnace. This makes the molten metal rise up the tube and enters the die cavity with low turbulence. After the metal has solidified, the air pressure is released. This makes the metal still in the molten state in the riser tube to fall back into the furnace. After subsequent cooling, the die is opened and the casting extracted.

With correct die design it is possible to eliminate the need of the riser also. This is because of the directional freezing of the casting. After the sequence has been established, the process can be controlled automatically using temperature and pressure controllers to oversee the operation of more than one diecasting machine. Casting yield is exceptionally high as there is usually only one ingate and no feeders.

Application of Pressure Die Casting

- Automotive parts like wheels, blocks, cylinder heads, manifolds etc.

- Aerospace castings.

- Electric motor housings.

- Kitchen ware such as pressure cooker.

- Cabinets for the electronics industry.

- General hardware appliances, pump parts, plumbing parts.

Cold Die Casting

Cold chamber die casting is the second of the two major branches of the die casting manufacturing process.

Cold Chamber Process

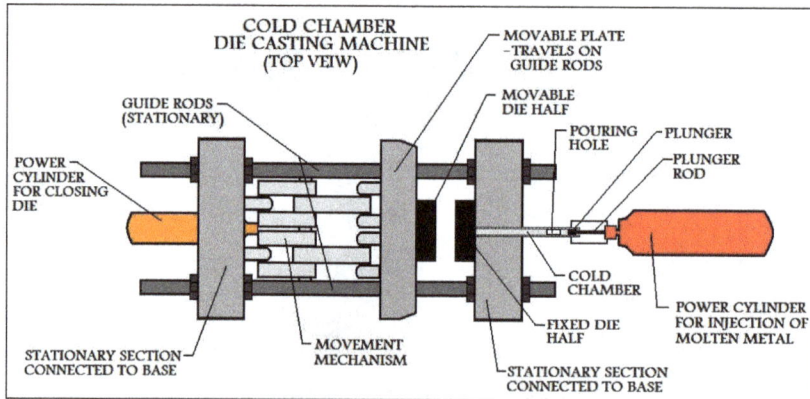

Cold chamber die casting machine (top view).

Cold chamber die casting is a permanent mold metal casting process. A reusable mold, gating system and all, is employed. It is most likely machined precisely from two steel blocks. Large robust machines are used to exert the great clamping force necessary to hold the two halves of the mold together against the tremendous pressures exerted during the manufacturing process.

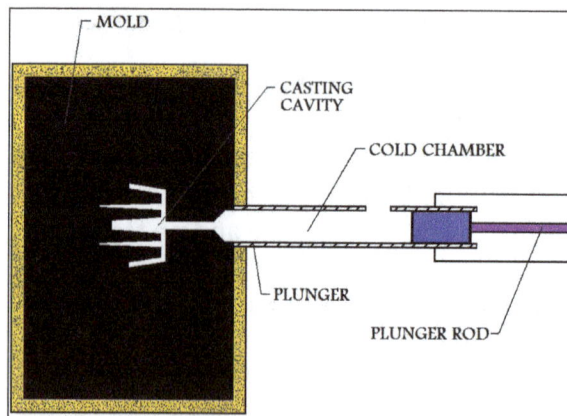

Cold chamber die casting.

A metal shot chamber, (cold-chamber), is located at the entrance of the mold. A piston is connected to this chamber, which in turn is connected to a power cylinder.

At the start of the manufacturing cycle, the correct amount of molten material for a single shot is poured into the shot chamber from an external source holding the material for the metal casting.

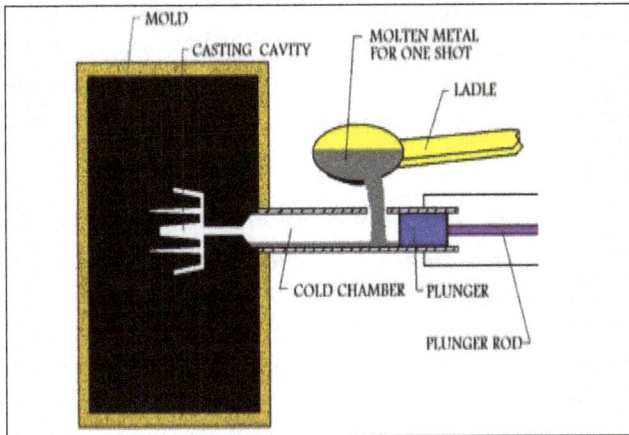

Cold chamber die casting.

The power cylinder forces the piston forward in the chamber, cutting off the intake port. The power cylinder moving the piston forward forces the molten material into the casting mold with great pressure. Pressure causes the liquid metal to fill in even thin sections of the metal casting and press the mold walls for great surface detail. The pressure is maintained some time after the injection phase of die casting manufacture.

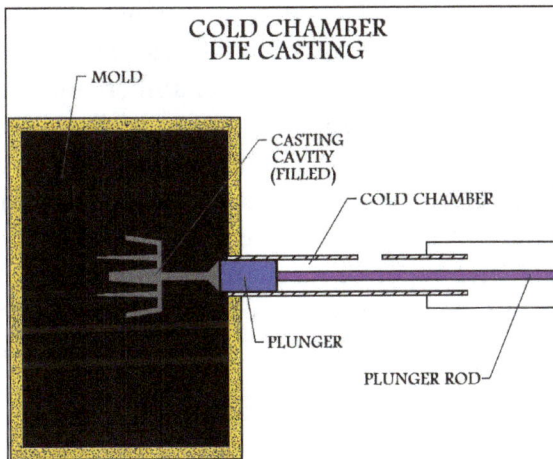

Cold chamber die casting.

Once the metal casting begins to solidify, the pressure is released. Then the mold is opened and the casting is removed by way of ejector pins. The mold is sprayed with lubricant before closing again, and the piston is withdrawn in the shot chamber for the next cycle of production.

Cold Chamber Die Casting for Manufacture

The main difference between cold-chamber die casting and hot-chamber die casting manufacture is that in the cold-chamber process the molten metal for the casting is introduced to the shot chamber from an external source, while in the hot chamber process the source of molten material is attached to the machine. In the hot-chamber process, certain machine apparatus is always in contact with molten metal. For this reason, higher melting point materials will create a problem for the machinery in a hot-chamber metal casting setup. Since the liquid metal is brought in from an outside source, the die casting machinery is able to stay much cooler in a cold-chamber process.

Consequently, higher melting point alloys of aluminum, brass, copper, and aluminum-zinc are often metal cast in manufacturing industry using cold chamber die casting. It is very possible to manufacture castings from lower melting point alloys using the cold-chamber method. When considering industrial metal casting manufacture, however, the advantages of production by the hot-chamber process usually make it the more suitable choice for lower melting point alloys.

In the cold chamber die casting process, material must be brought in for every shot or cycle of production. This slows down the production rate for metal casting manufacture. Where in the hot chamber process, castings can be constantly output. Cold chamber die casting should still be considered a high production manufacturing process.

In comparison with the hot die casting process, the cold die casting process requires the application of more pressure. The pressure at which the molten metal is forced into and fills the die cavity in cold chamber metal casting manufacture typically outranks the pressure used to fill the die in hot chamber metal casting by about an order of magnitude. Pressures of 3000psi to 50000psi (20MPa to 350MPa) may be used in manufacturing industry to fill the mold cavities with molten material during cold chamber die casting manufacture. Castings manufactured by cold chamber die casting have all the advantages characteristic of the die casting process, such as intricate detail, thin walls, and superior mechanical properties. The significant initial investment into this manufacturing process makes it suitable for high production applications.

Gravity Casting

Gravity casting is a casting process used for non-ferrous alloy parts. Sometimes referred to as Permanent Mould, the process is typically used on aluminum, zinc, and copper base alloys.

Process Gravity Casting

There are three stages to the gravity casting process:

- The first step is the heating of the mold and coating it with a die release agent. The release agent spray also serves as a cooling agent after the part has been removed from the die.

- In the second step, the molten metal is poured into channels in the tool to allow the material to fill the entire mold cavity.

- In third step, the metal is dosed and hand poured by using ladles. Usually, there is a mold "down sprue" that allows the alloy to enter the mold cavity from the lower part of the die. This reduces the formation of turbulence and subsequent porosity and inclusions in the finished part. After the part has cooled, the die is opened by either using a mechanical tool or manually.

Advantages

There are many advantages to gravity casting. It offers good dimensional accuracy, a smoother cast surface finish than sand casting, and improved mechanical properties compared to sand casting. Gravity casting also provides faster production times compared to the other processes.

Centrifugal Casting

Centrifugal casting is one of the largest casting branches in the casting industry, accounting for 15% of the total casting output of the world in terms of tonnage. Centrifugal casting was invented in 1918 by the Brazilian Dimitri Sensaud deLavaud, after whom the process was named. DeLavaud's invention eliminated the need for a central core in the pipe mold, and the mold was water cooled, allowing for a high rate of repeated use. The technique uses the centrifugal force generated by a rotating cylindrical mold to throw molten metal against a mold wall to form the desired shape. Therefore, a centrifugal casting machine must be able to spin a mold, receive molten metal, and let the metal solidify and cool in the mold in a carefully controlled manner.

All metals that can be cast by static casting can be cast by the centrifugal casting process, including carbon and alloy steels, high-alloy corrosion- and heat-resistant steels, gray iron, ductile and nodular iron, high-alloy irons, stainless steels, nickel steels, aluminum alloys, copper alloys, magnesium alloys, nickel- and cobalt-base alloys, and titanium alloys. Non-metals can also be cast by centrifugal casting, including ceramics, glasses, plastics, and virtually any material that can be made into liquid or pourable slurries. Centrifugal castings can be best described as isotropic, that is, having equal properties in all directions. This is not true of a forging, rolling, or extrusion.

The centrifugal technique is used primarily for the production of hollow components, but centrifugal casting is used to create solid parts. The centrifugal casting process is generally preferred for producing a superior-quality tubular or cylindrical casting, because the process is economical with regard to casting yield, cleaning room cost, and mold cost. The centrifugal force causes high pressures to develop in the metal, and it contributes to the feeding of the metal, with separation from nonmetallic inclusions and evolved gases. In centrifugal casting of hollow sections, nonmetallic inclusions and evolved gases tend toward the inner surface of the hollow casting. By using the outstanding advantage created by the centrifugal force of rotating molds, castings of high quality and integrity can be produced because of their high density and freedom from oxides, gases, and other nonmetallic inclusions. When casting solid parts, the pressure from rotation allows thinner details to be cast, making surface details of the metal-cast components more prominent. Another advantage of centrifugal casting is the elimination or minimization of gates and risers.

Centrifugal Casting Methods

Centrifugal casting machines are categorized into three basic types based on the direction of the spinning axis: Horizontal, vertical, or inclined. Centrifugal casting processes also have three types:

- True centrifugal casting (horizontal, vertical, or inclined).

- Semicentrifugal (centrifugal mold) casting.

- Centrifuge mold (centrifugal die) casting.

The latter two methods are only done with vertical spinning.

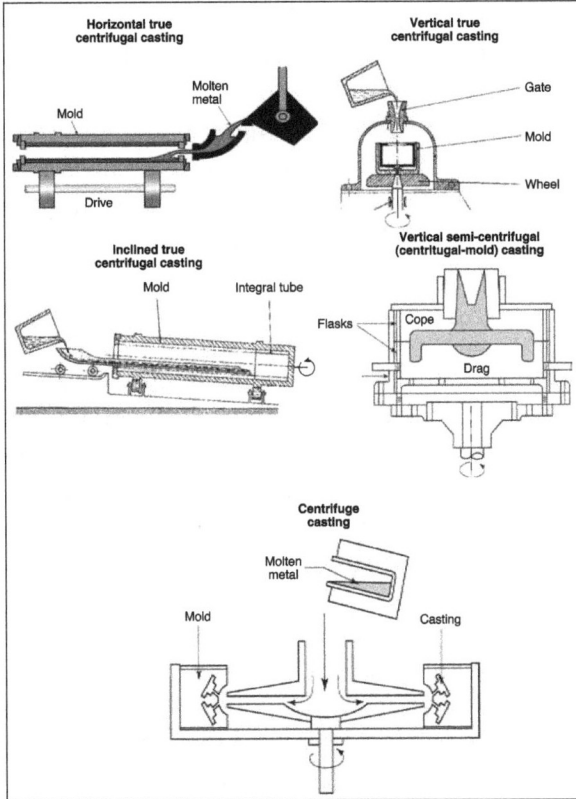

Methods of casting with rotating molds. True centrifugal casting (top row) can be horizontal, vertical, or at an incline. Semicentrifugal and centrifuge die casting are vertical methods with molds that can be designed to produce solid parts.

Horizontal centrifugal casting is mainly used to cast pieces with a high length-to-diameter ratio or with a uniform internal diameter. Products include pipe, tubes, bushings, cylinder sleeves (liners), and cylindrical or tubular castings that are simple in shape. On the other hand, vertical centrifugal casting is mainly for castings with a low length-to-diameter ratio (except vertically cast extralong rolls) or with a conical diameter. The product range for vertical centrifugal casting machines is wider, because noncylindrical (or even nonsymmetrical) parts can be made using vertical centrifugal casting. All vertical centrifugal castings have more or less taper on their inside diameters, depending on the gravitational (g) force applied to the mold and the casting size. The inclined centrifugal casting machine bears advantages and disadvantages of both horizontal and vertical castings and can be very useful in certain applications.

Although both vertical and horizontal methods employ centrifugal force, there are some differences in how the force is applied with respect to the axis of the mold rotation and the speed of the molten metal relative to the rotating mold. For example, with a vertical mold axis, the resultant force on the liquid is constant. This is not the case in a horizontal mold. The other difference between horizontal and vertical mold orientation is the speed obtained by the molten metal as it spins around the mold. When metal is poured into the horizontally rotating mold, considerable slip occurs between the metal and the mold such that the metal does not move as fast as the rotating mold. To overcome this inertia, the metal must be accelerated to reach the mold rotation speed. This is not a problem in the vertical centrifugal process, where the molten metal reaches the speed of the mold soon after pouring. However, with a vertical mold axis, there is a tendency for the molten metal to form a parabolic shape due to the competing gravitational and centrifugal forces.

True Centrifugal Casting

This method, also referred to as just centrifugal casting, is characterized by an outer cylindrical mold with no cores. The process can be vertical, horizontal, or inclined. The permanent mold is rotated about its axis at high speeds (300 to 3000 rpm), so that the molten metal is forced to the inside mold wall, where it solidifies. The casting is usually very fine grained on the outer diameter, while the inside diameter has more impurities and inclusions that can be machined away.

Centrifugal casting is used to produce cylindrical, tubular, or ring-shaped castings. The need for a center core is completely eliminated. Castings produced by this method will always have a true cylindrical bore or inside diameter, regardless of shape or configuration. The bore of the casting will be straight or tapered, depending on the horizontal or vertical spinning axis used. Castings produced in metal molds by this method have true directional heat flow, facilitating a planar solidification front move from the outside of the casting toward the axis of rotation. This method results in the production of high-quality, defect-free castings without shrinkage, which is the largest single cause of defective sand castings.

Centrifugal casting is also the preferred method for Babbitting medium- and thick-wall, half-shell or full-round (nonsplit) journal bearings, because it virtually eliminates porosity and allows close control of the cooling process to promote a strong bond. A disadvantage is the need for more extensive equipment and tooling than for static casting. Minor segregation of the intermetallics can occur across the thickness of the Babbitt, but segregation along the axial length of a statically cast bearing can be more serious and is more difficult to detect. The spinning axis is usually horizontal, but vertical orientation is sometimes employed for unusual sizes (e.g., large diameters or short lengths).

Advantages of centrifugal casting include:

- Flexibility in casting composition: Centrifugal casting is applicable to nearly all compositions, with the exception of high-carbon steels (0.40 to 0.85% C). Carbon segregation can be a problem in this composition range.

- Wide range of available product characteristics: The metallurgical characteristics of a tubular product are mainly characterized by its soundness, texture, structure, and mechanical properties. Centrifugal castings can be manufactured with a wide range of microstructures tailored to meet the demands of specific applications.

- Dimensional flexibility: Horizontal centrifugal casting allows the manufacture of pipes with maximum outside diameters close to 1.6 m (63 in.) and wall thicknesses to 200 mm (8 in.). Tolerances depend on part size and the type of mold used.

- Quality: The centrifugal action removes unwanted inclusions, dross, cleaner casting, and material that contain shrinkage, which can be machined away. Class I castings can be produced without the need for upgrading and costly weld repairs.

- Properties: Mechanical properties are often superior to those of static castings due to the finer grains resulting from the process, which are of constant size in circumferential and axial directions. Due to cleanliness and finer grain size, good weldability is achieved.

Semicentrifugal (Centrifugal Mold) Casting

In semicentrifugal casting, a mold is rotated around its axis of symmetry. Cast configurations may be complex, determined by the shape of the mold. Molds for semicentrifugal casting often contain cores for production of internal surfaces. Directional solidification is obtained only by proper gating, as in static casting. Castings that are difficult to produce statically can often be economically produced by this method, because centrifugal force feeds the molten metal under pressure many times higher than that in static casting. This improves casting yield significantly (85 to 95%), completely fills mold cavities, and results in a high-quality casting free of voids and porosity. Thinner casting sections can be produced with this method than with static casting. Typical castings of this type include gear blanks, pulley sheaves, wheels, impellers, cogwheels, and electric motor rotors. The centrifugal force is used for slag separation and refilling of melt.

Centrifuge (Centrifugal Die) Casting

Centrifugal mold or die casting, also referred to as centrifuge casting, has the widest field of application. This casting method is typically used to produce valve bodies and bonnets, plugs, yokes, brackets, and a wide variety of various industrial castings.

In this method, the casting cavities are arranged about the center axis of rotation like the spokes of a wheel, thus permitting the production of multiple castings. Centrifugal force provides the necessary pressure on the molten metal in the same manner as in semicentrifugal casting. The centrifugal force contributes to metal feed for casting thinner sections and making surface details.

Centrifuge casting is often used in conjunction with investment casting. It is also used with rubber molds made of silicone rubber with sufficiently higher temperature resistance for repeated use in casting without mold deterioration. Centrifuge casting with rubber molds is also referred to as spin casting. Suitable materials for spin casting include zinc-base alloys, lead-base alloys, tin-base alloys, aluminum, and plastics.

Equipment

A centrifugal casting machine must be able to perform six operations accurately and with repeatability:

- The machine must be able to accelerate the mold to a predetermined speed, maintain smooth spinning, and decelerate to a stop in a reasonable time frame. The machine also should be able to change speeds during pour and solidification for some special applications.

- There must be a way to heat and coat the mold before pouring the molten metal (except the cold mold method for ductile iron pipe).

- There must be a means to pour the molten metal safely into the rotating mold at a controlled rate, position, and orientation.

- Once the metal is poured, a proper solidification and cooling rate must be established in the mold to obtain a desired casting microstructure, protect the mold, and achieve required productivity.

- There must be a means of adding inoculants or fluxes for some special applications.

- There must be a means of extracting the solidified casting quickly from the mold at elevated temperatures without deforming the casting.

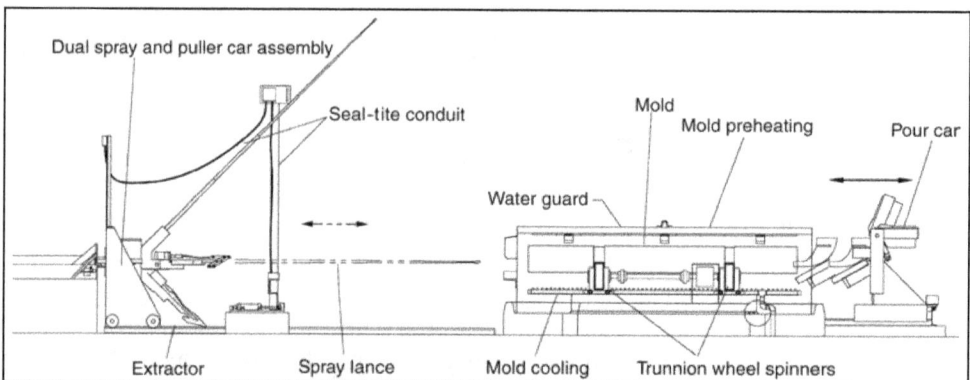

Automated horizontal casting machine.

Centrifugal casting machines have realized mechanization and automation basically in all of the aforementioned six tooling areas in many applications. An example of an automated horizontal machine and extractor is shown in figure. New casting and machine products are still evolving, such as metal-matrix composite castings and magnetic stirring solidification. Centrifugal casting and equipment engineers continue to use computers to collect, store, and process the key production, equipment, and product information for high-tech products, so that the production parameters can be optimized and automatically set up in production. Centrifugal casting will continue to be prosperous in the casting industry.

Spinners

Spinning of the mold is realized by spinners. The achievable spinning speed and, sometimes, the ability to change speeds during the pour and solidification process is an important consideration for determining whether a casting geometry or a casting microstructure can be achieved. Spinning speeds are chosen based on the centrifugal force requirement, which is measured by the multiple of the gravitational force (called g force). Horizontal casting machines are typically spun at 45 to 60 g force on the casting inside diameter, and vertical casting machines are typically spun at 75 g force. For some special applications, the spinning speed can be as high as 100 to 200 g force (e.g., horizontal roll castings and heat-resistant steel tube castings); in other cases, it can be as low as a fraction of a g to a few g's (e.g., when vertically pouring a roll core, vertically pouring solid bars, or horizontal two-speed pouring). Figure in the article "Vertical Centrifugal Casting" in this Volume provides a convenient chart to check the centrifugal force against casting diameters and spinning speeds.

In order to achieve the correct g force, the proper motor size is required. Too small a motor cannot meet process requirements for the spinning speed, and too big a motor can have an excessive acceleration rate that may cause machine vibration. For large horizontal machines, hydraulic motors are preferred over electric motors, because the former offer smoother spinning. Vibration monitors are installed for some large spinners to ensure smooth spinning. The spinners also need to be stopped at a predetermined time, which can be realized by the proper electric, hydraulic, or mechanical means. Most vertical machines use a single spinner, and the mold is spun on top of the spinner. These vertical machines will have less stability than horizontal machines when the mold length-to-diameter ratio is greater than 1.

Vertical machines with a long mold need trunnion wheels to support the mold around the circumference at the upper section. Small spinners and molds can be clustered to form a turntable, an over-and-under, or a Ferris wheel-type casting machine, which shares the same pouring, heating, coating, and extracting systems, forming a very efficient production cell.

Pour Equipment

Pour equipment and pour rate can be essential pieces of the process for successful production of certain types of castings. Pouring is a critical factor in the following applications, to name a few. The centrifugal ductile iron pipe casting process requires a long, moving trough and quadrant pour ladle to deliver the molten iron along the mold to make 6 to 8 m (20 to 26 ft) long pipe. The centrifugal soil pipe casting process needs a proper chute that can be cleaned and coated in a few seconds after each pour and a consistent pour rate to dash the molten metal to the far end of the mold to form 3 m (10 ft) long pipes with even thickness. Centrifugal heat-resistant steel tube castings rely on a fast pour rate and sufficient pour height to help form the long thin-wall tube before the molten metal gets too cold. Vertical roll casting uses a special-shaped down sprue to prevent the metal flow from whirling.

Casting Cooling

Centrifugal castings should be cooled unidirectionally from outside to inside. Any two-way solidification will increase the chance of shrinkage porosity and machining allowance, which should be avoided or minimized in thick-wall castings. Cooling rate can affect the microstructure, casting hardness, circumferential and axial cracks, machine productivity, as well as mold life. In most cases, the early cooling rate of the castings is mainly controlled by the coating thickness, coating texture, coating materials, as well as mold thickness and mold materials; however, the later cooling rate is mainly controlled by water cooling (unless the mold is not cooled by water). Water-cooling methods include water submerge, waterjet spray, and water sleeve. Ductile iron pipe production uses all three water-cooling methods mentioned previously, but other centrifugal castings mainly require waterjet spray, because of its simple setup and easy control of mold temperature. Roll production usually does not require water cooling, because the chill wall thickness is sufficient to absorb heat from the molten metal. Water cooling needs to be consistent, but sometimes, it needs intensity variation along the casting length. For example, the middle section of a long tube mold usually needs more water for cooling.

As far as the grain structure of centrifugal castings (e.g., columnar versus equiaxed) is concerned, the pour temperature or the variable spinning speed plays a much more important role in obtaining the equiaxed grains than the water-cooling rate or mold temperature. Water cooling can be critical in certain products, such as high leaded bronze bushings and Babbitt bearings, where intensive water cooling assures that the lead segregation is suppressed.

Some applications require continuous monitoring of casting and mold temperatures by a thermal couple probe inserted into the mold cavity or infrared sensors aimed at the casting inside diameter and mold outside diameter, so that the current process can be stopped and the next process can be started. This is critical, for example, to obtain a good bimetal bonding for roll production.

Casting Inoculation and Fluxing

Some castings necessitate the use of inoculants to obtain the desired microstructure (such as ductile iron and gray iron) or grain sizes (such as thick-wall chilled iron or steel tubes). In some applications, fluxes are needed to protect the casting inside diameter in order to obtain a sound metallurgical bimetal bonding (such as in the production of rolls and brake drums) or to prevent premature solidification or oxidation (such as in thick-wall steel castings) of the casting inside diameter. Still others may use fluxes to purify the molten metal during pouring and solidification. All of these would require some means to deliver the inoculants or fluxes into the metal stream during pouring or into the mold cavity before or during solidification.

Casting Extraction

Castings need to be extracted from the permanent mold. Small and low-production machines use manual extractors and manpower to extract. Large castings, however, rely on mechanized extractors to do the job. In extracting from horizontal machines, the extractor must overcome the friction acting on the whole interface between the casting and the mold. For vertical machines, the extractor must overcome friction plus casting weight. For some applications, the castings are kept rotating during and/or after extraction to prevent the casting from warping.

Molds

Sand molds, semipermanent molds, and permanent molds can be used for the centrifugal casting process. Centrifugal action can also be combined with other molding methods such as investment casting. Selection of the type of mold is determined by the shape of the casting, the degree of quality needed, material to be cast, and the production (number of castings) required. In addition to the processing parameters, proper mold design and selection is vital to producing quality castings. Molds consist basically of four parts: the mold body, track grooves or roller tracks on the body (these can be absent if the machine has thrust wheels to hold the mold axially), endplates attached to the mold body, and endplate swing locks or taper pins/wedges. A vertical mold needs fixtures to fasten it onto the adapter table, and the table is bolted onto the spinner shaft. A mold used on a dual-faceplate horizontal machine can have a mold body without other mold parts.

Mold materials include metallic permanent molds, refractory-lined metal molds, sand-lined metal molds, and other materials such as graphite and rubbers. The metallic permanent molds are most widely used because of their reusability, accurate casting geometry, and high productivity. Mold inside diameters are subject to thermal fatigue no matter what mold material is used. The low-carbon and low-alloyed forged steel molds have much longer fatigue lives than cast steel or cast iron molds, and they are also safer to use. However, forged steel molds are more expensive to make than molds made of cast steel and iron. Copper alloys, tool steels, and superalloys are sometimes used for small castings that require high pour temperatures or high cooling rates. Hardened rubber molds are used for some metal jewelry with low melting points.

Both horizontal and vertical molds must meet certain geometrical precisions (straightness, smoothness, roundness, concentricity, freedom from internal pores) to minimize machine vibration, which can be critical to safety, bearing life, machine life, and product quality. The mold should also be maintained or repaired during the life of its service. For example, the inside diameter of a ductile iron pipe mold is periodically peened to eliminate tensile stress caused by thermal fatigue. Large grooves or cracks on the mold inside diameter can be repaired by subarc welding and grinding. The mold geometric accuracy can be maintained through remachining and grinding.

Mold endplates should be properly designed to prevent molten metal leakage. It cannot be overemphasized that the mold endplate locks and mold fasteners are designed with a safety factor greater than 10 to prevent the endplates from coming off the mold or the molds from coming off the machine, which can cause catastrophic molten metal spilling. The push force (F) on the endplates generated by the spinning metal is expressed by the following equation:

$$F = 5.375\rho\left(\frac{N}{100}\right)^2 \left(d_o^2 - d_i^2\right)^2$$

where ρ is the molten metal density (kg/m³), N is the mold spinning speed (rpm), d_o is the casting outside diameter (m), and d_i is the casting inside diameter (m).

This equation illustrates that the pushing force increases rapidly with the spinning speed (squared), casting diameter (squared), casting thickness (squared), as well as the metal density. Therefore, whenever a large or thick casting is to be poured, the pushing force on the endplates must be calculated so that the stresses of the endplate locks, bolts, pins, and fasteners can be calculated and properly designed.

Mold Heating and Coating Techniques

The centrifugal casting mold must be heated and coated with ceramic mold wash, mold powder, sand, resin sand, or graphite. The most widely used mold coating is water-based mold wash, which is consistently replacing the other coating materials. Sand and resin sand linings are quickly fading out, and graphite coating is mainly used for small nonferrous castings. The dominance of water-based mold wash is established on the fact that it offers significant advantages over the other types. Some of these advantages are:

- Water-based mold wash is insulating enough for almost all applications.

- It allows immediate metal pouring as soon as the coating application is finished, which greatly increases the casting productivity.

- It offers a better surface finish to the castings.

- Coating thickness can be easily used to control the microstructure.

- It greatly reduces the friction between the casting and the mold.

- It can be easily cleaned off the casting and mold.

- Its consumption is very small compared to sand linings.

- The mold has a long life under its protection.

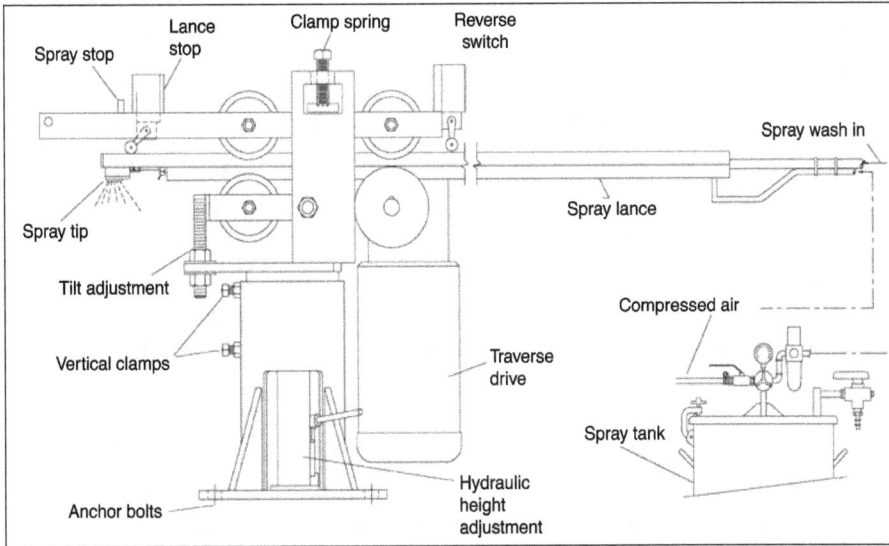

Patented automatic mold wash spray assembly.

The mold can be heated from the outside or inside diameter on the machine with the mold spinning slowly by a row, or rows, of gas burners or off the machine in an oven at 180 to 320 °C (355 to 610 °F). The mold temperature can be critical for the coating adhesion to the mold, coating strength, as well as casting surface quality. Once the mold is heated sufficiently, mold wash is applied on the mold inside diameter by spraying or flooding. The mold wash thickness is usually between 0.5 and 3 mm (0.02 and 0.12 in.), depending on the application. The spraying method is preferred for most processes because it gives more uniform coating thickness, smoother coating finish, and more consistent coating quality. Spraying can also reach the areas that flooding cannot. The spraying method works for both horizontal and vertical molds, but the flooding method works only for horizontal molds. Figure shows a patented automatic spray lance system that offers automatic control of the forward and backward movement, moving velocity, and spray on and off operations of the spray lance. It can also adjust height and tilt angle to align the lance with all customer mold sizes.

Defects in Centrifugal Castings

The three most common defects observed in centrifugal castings are segregation banding, raining, and vibration defects:

- Segregation banding occurs only in true centrifugal casting, generally where the casting wall thickness exceeds 50 to 75 mm (2 to 3 in.). It rarely occurs in thinner-wall castings. Banding can occur in both horizontal and vertical centrifugal castings. Banding is more prevalent in alloys with a wide solidification range and greater solidification shrinkage.

Bands are annular segregated zones of lowmelting constituents, such as eutectic phases and oxide or sulfide inclusions. They are characterized by a hard demarcation line at the outside edge of the band that usually merges into the base metal of the casting.

Most alloys are susceptible to banding, but the wider the solidification range and the greater the solidification shrinkage, the more pronounced the effects may be. Banding has been found when some critical level of rotational speed is attained, and it has been associated with very low speeds, which can produce sporadic surging of molten metal. Therefore, both mechanisms may be involved. Minor adjustments to casting operation variables, such as rotational speed, pouring rate, and metal and mold temperatures, will usually reduce or eliminate banding.

- Raining is a phenomenon that occurs in horizontal centrifugal castings. If the mold is rotated at too low a speed or if the metal is poured into the mold too fast, the metal actually rains or falls from the top of the mold to the bottom. Proper process control can eliminate raining.

- Vibration defects can cause a laminated casting. It can be held to a minimum by proper mounting, careful balancing of the molds, and frequent inspection of rollers, bearings, and other equipment.

Applications

Typical products produced by horizontal centrifugal casting machines include pipe and tubes of all different metals; alloyed iron, steel, and high-speed steel rolls for steel mills and the food-processing, papermaking, and printing industries; iron and steel cylinder liners, gray iron piston rings, gray iron brake drums, and alloy and superalloy seal rings and valve seats for the auto industry; heat-resistant steel cracking tubes for petroleum refinery and radiant heating tubes and furnace rolls for heat treatment furnaces; and copper alloys and Babbitt alloys for bimetal bearings and bushings.

Vertical machines are mainly used for shorter parts, such as bushings, gear blanks, rings, short rolls, wheels, aluminum and copper electric motor rotors, jewelries, and vacuum titanium alloy parts, and many different small round parts, such as balls, valve bodies, and rings for the machinery industry. With proper support, vertical casting machines can also produce extralong rolls. Slightly inclined machines are used for producing ductile iron pipe for water and gas mains, and other inclined machines are used for rolls and conical bushings.

Centrifugal casting is also used to ensure good filling in investment casting, as described subsequently. Another related technique is spin casting, which uses rubber molds for casting of zinc, tin, and other alloys with low melting points. It is also used in conjunction with nonmetallic materials such as glass and plastics. A more advanced materials application is the combustion synthesis of functionally graded materials.

Investment Casting with Centrifugal Force

When the action of centrifugal force is combined with investment casting, the main benefit is an improvement in casting soundness with subsequent improvements in mechanical properties. The influence of centrifugal force in aluminum investment castings found that:

- A down-tapered sprue was more effective in eliminating turbulence of the metal stream and nonmetallic incisions.

- Surface smoothness and dimensional accuracy of aluminum alloy centrifugal castings were much better than the gravity investment castings, but they were much more affected by the fineness of investment material than centrifugal force.

- The soundness and mechanical properties were improved with increased centrifugal force.

The liquidus-solidus interval is important in centrifugally cast investment mold castings, and final casting soundness depends on the gating system for castings of alloys with narrow and wide freezing ranges. Factors that influence soundness were identified to be:

- Ingate diameter.

- Sprue diameter.

- Metal pour temperature.

- Mold temperature prior to pour.

- Mold speed.

- Spinning period.

- Casting weight (expressed as a casting length, l).

- Number of castings in each ring on the sprue.

- Distance (L) from the axis of rotation to the castings on the sprue (which characterizes the sprue length).

- Liquidus-solidus interval (freezing range).

For a wide freezing range (e.g., 100 K in bronze), the casting soundness was improved by increasing mold speed, ingate diameter, sprue diameter, and the product (l × L) of specimen size and sprue length. Mold speed, n, was found to be the most significant variable.

For wide-interval alloys, optimal conditions are "those under which each portion of molten metal newly arriving in the mold cavity should spread out at the appropriate level and freeze onto the solidification front without building up a substantial surplus of molten metal. Thus, the optimum conditions for pouring intricate castings in wide-interval alloys are those which produce layer-by-layer freezing."

"Casting soundness decreased as the number of castings at the same level on the sprue increased. Thus, as fewer castings are made, their density will increase. Increasing the number of ingates surrounding the sprue at the same level probably leads to the development of a hot spot, heat storage around the spot, and departure from conditions required for layerwise solidification. It is therefore best to assemble the patterns in a helical array on the sprue for intricate castings in wide-interval alloys."

"Density comparisons between the centrifugal castings at the two extreme levels on the sprue have shown that the sprue length has no significant influence on density." For narrow interval alloys, there is a sprue-length effect. Narrow-interval alloys require much greater ingate and sprue cross sections than wide-interval alloys, and narrow-interval alloy "castings should always be made on short sprues."

"Narrow-interval alloys require heat accumulation around the sprue and ingates. Sound castings can be made by accumulating surplus molten metal ahead of the solidification front. In this case, the optimum conditions correspond to wide-section ingates and sprues, mold preheating, the use of short sprues, and some reduction in the relative rate of molten metal supply to the mold cavities."

"Intricate castings in wide-interval alloys must be made under conditions which will prevent excessive heat accumulation around the gating system and mold cavities, and minimize the volume of molten metal ahead of the solidification front. Sound metal can be ensured by directional solidification. In this case, mold preheating, commensurate metal weights and sprue lengths, helical mold assemblies around the sprue, and normal ingate and sprue cross sections should lead to a certain 'freezing' action, while rapid mold cavity filling with metal in this condition should lead to layerwise solidification".

Centrifugal Casting with Combustion Synthesis

Combustion synthesis (CS) is an attractive technique for synthesizing a wide variety of advanced materials, including powders and near-net-shape products of ceramics, intermetallics, composites, and functionally graded materials. One type of CS is self-propagating high-temperature synthesis (SHS). Combustion synthesis of highly exothermic reactions (typically reduction type) often results in completely molten products, which may be processed using common metallurgical methods.

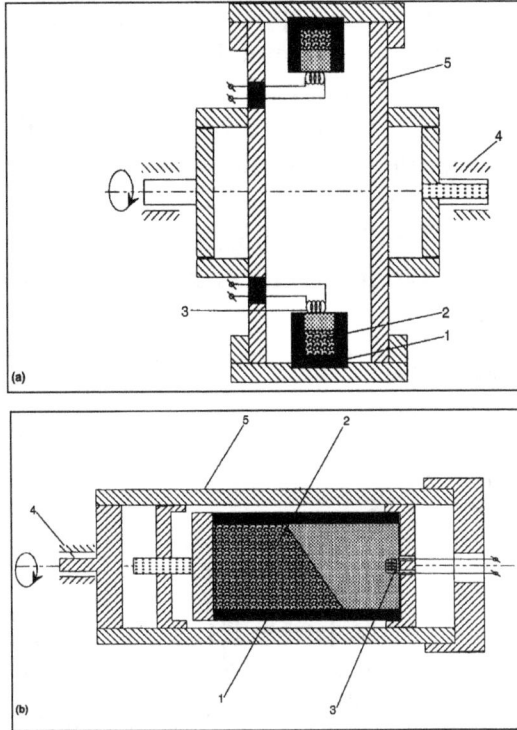

Schematic of self-propagating high-temperature synthesis plus centrifugal casting. (a) Radial centrifuge. (b) Axial centrifuge. 1, sample container; 2, reactant mixture; 3, ignitor; 4, axle; 5, reactor.

Casting of CS products under inert gas pressure or centrifugal casting has been used to synthesize cermet ingots, corrosion- and wear-resistant coatings, and ceramic-lined pipes. Casting under gas pressure is similar to the conventional SHS production, while centrifugal casting allows for greater control of the distribution of phases by controlling the time of separation. For radial centrifuges, the sample is placed at a fixed radial position from the axis of rotation, and the applied centrifugal force is parallel to the direction of propagation. The influence of centrifugal acceleration characterizes the phase distribution in the final product.

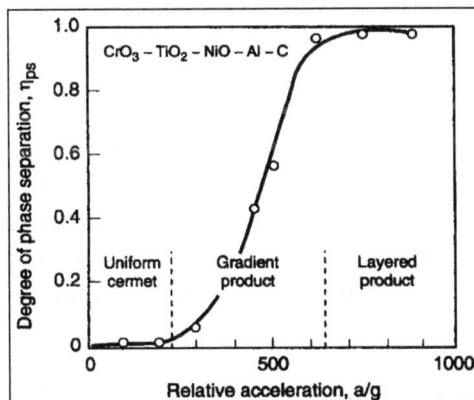

Degree of phase separation as a function of centrifugal acceleration, a, where g is acceleration due to gravity.

The second type of centrifugal casting apparatus, called an axial centrifuge, is used for production of ceramic, cermet, or ceramic-lined pipes. The axial centrifuge casting method was developed further for production of long pipes with multilayer ceramic inner coatings. In this process, a thermite mixture (e.g., $Fe_2O_3/2Al$) is placed inside a rotating pipe and ignited locally. A reduction-type combustion reaction propagates through the mixture, and the centrifugal force results in separate layers of metal and ceramic oxide, with the latter forming the innermost layer. The process is carried out in air under normal pressure, and pipes with diameters up to 30 cm (12 in.) and 5.5 m (18 ft) long have been obtained.

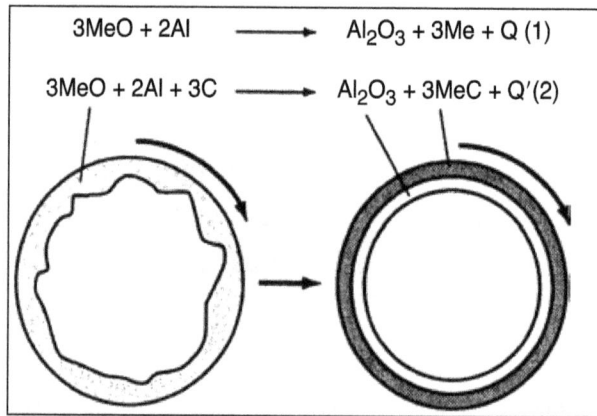

$$3MeO + 2Al \longrightarrow Al_2O_3 + 3Me + Q \ (1)$$

$$3MeO + 2Al + 3C \longrightarrow Al_2O_3 + 3MeC + Q' \ (2)$$

Concept of centrifugal process for production of ceramic-lined steel pipes.

Spin casting typically refers to a special type of centrifuge casting with rubber molds. In the early 1970s, heat-cured silicone rubber (General Electric Silcast) was introduced with a sufficiently higher temperature resistance for molds in casting plastics and metals with low melting points, such as zinc alloys, tin alloys, and aluminum. Heat-cured silicone rubber molds can withstand temperatures up to approximately 550 °C (1020 °F) at a rate of 50 to 60 cycles per hour for hundreds of cycles. However, spin casting of metals with rubber molds is generally limited to the low-melting-point alloys (below 400 °C, or 750 °F), due to the degradation of the silicone rubber. Mold life is on the order of 200 to 250 shots at 400 °C (750 °F).

During pouring, the mold is placed on a centrifuge wheel to achieve adequate metal filling into intricate areas of the mold. Depending on the size of the mold and the material being cast, spinning speeds range from 100 to 900 rpm. Metal products can range from less than 100 g to 1.2 kg (3.5 oz to 2.6 lb). Maximum product size depends on mold diameter and thickness. Mold diameters can range up to approximately 60 cm (24 in.), with thicknesses of 10 cm (4 in.) or more. In mold making, metal patterns are laid on uncured silicone rubber sheets, which are packed into a circular metal frame. The circular metal frame is placed into a hydraulic vulcanizer press and electrically heated for curing.

The curing temperature and duration depends on the type of rubber and the mold thickness. Pattern materials, such as pewter, can be used as long as they can withstand a temperature of 190 °C (375 °F) under several thousands of pounds pressure without distorting, melting, breaking, or outgassing. The heat melts the rubber, which is pressed against the surface of the pattern before the rubber sets. After vulcanization, the mold is separated from the metal, and the patterns are removed. Special-shaped tools are used to cut gates and runners on the rubber. Air vents are also cut for air to escape from the mold cavity. Modifications to gates and vents are usually required after a few shots to obtain the best results. Making of the mold takes approximately 1 to 3 days, depending on the complexity of the product shape.

While the mold is spinning, the liquid metal or plastic is melted in a suitable electric or gasfired crucible furnace and then poured into the center sprue of the mold. After the metal solidifies (the plastic parts set up), the parts are quickly removed from the mold. With metal, 50 to 60 cycles per hour can readily be made; with plastic, 10 to 15 cycles are typical. Parts with undercuts can be demolded from the rubber mold with no difficulty. Thin-walled parts can be molded comparable to die casting. The rubber mold is not as precise as the metal mold, but the surface finish is generally good. However, blisters (very small pinpoint holes) were often found on the surface of the thick section due to the shrinkage of metals. The blisters are the defects for products with a plated glossy surface.

Metal Products

Spin casting has the capability to produce intricate designs for low- or high-volume quantities. Metal products include:

- Cams and levers in automotive control panels.

- Bushings.

- Hub, clamps, and screws.

- Triggers, firing mechanisms, and cartridges.

- Medals, pins, figurines, key chains, name plates, and costume jewelry.

- Door hardware, locks, hooks, and rings.

- Handles.

- Belt buckles, buttons, and watch cases.

- Gears, motor housings, and pump impellers.

Plastic Products

Spin casting of thermoplastics and thermoset plastics includes:

- Plumbing fittings, valves, and couplings.

- Electrical switch plates, condensers, and connector plugs.

- Toy cars, wheels, and gears.

- Pen holders, clips, and fasteners.

- Electronic computer parts.

- Calculator casing keys.

Wax patterns for investment casting are also produced by spin casting.

True Centrifugal Casting

The manufacturing process of centrifugal casting is a metal casting technique that uses the forces generated by centripetal acceleration to distribute the molten material in the mold. Centrifugal casting has many applications in manufacturing industry today. The process has several very specific advantages. Cast parts manufactured in industry include various pipes and tubes, such as sewage pipes, gas pipes, and water supply lines, also bushings, rings, the liner for engine cylinders, brake drums, and street lamp posts. The molds used in true centrifugal casting manufacture are round, and are typically made of iron, steel, or graphite. Some sort of refractory lining or sand may be used for the inner surface of the mold.

Process

It is necessary when manufacturing a cast part by the true centrifugal metal casting process, using some mechanical means, to rotate the mold. When this process is used for industrial manufacture, this is accomplished by the use of rollers. The mold is rotated about its axis at a predetermined speed. Molds for smaller parts may be rotated about a vertical axis. However, most times in true centrifugal casting manufacture the mold will be rotated about a horizontal axis. The effects of gravity on the material during the metal casting process make it particularly necessary to cast longer parts with forces generated from horizontal rather than vertical rotation.

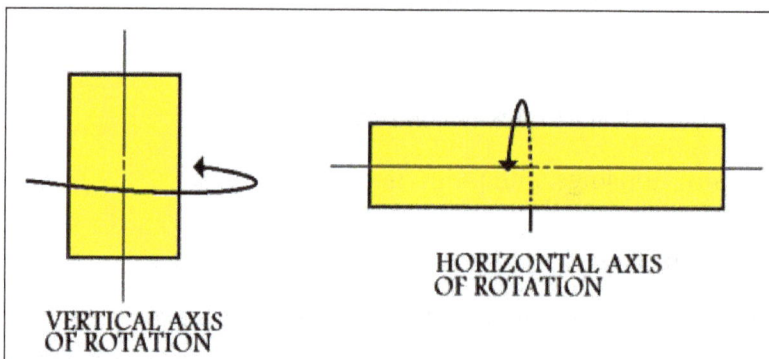

Vertical vs. Horizontal rotation.

The molten material for the cast part is introduced to the mold from an external source, usually by means of some spout. The liquid metal flows down into the mold. Once inside the cavity, the centripetal forces from the spinning mold force the molten material to the outer wall. Molten material for the casting may be poured into a spinning mold or the rotation of the mold may begin after pouring has occurred.

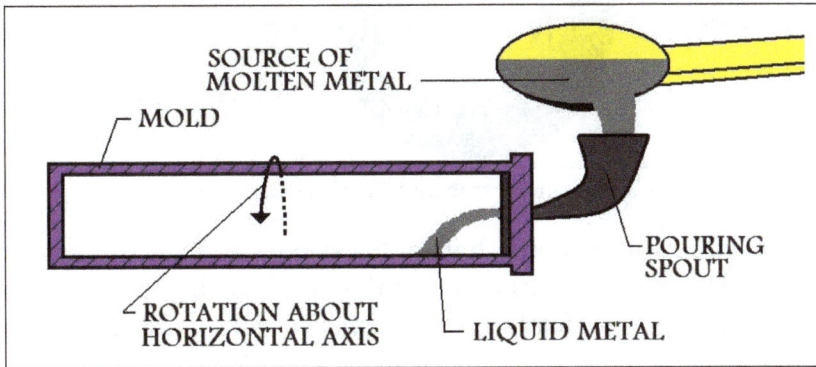

Pouring in true centrifugal casting.

The metal casting will harden as the mold continues to rotate.

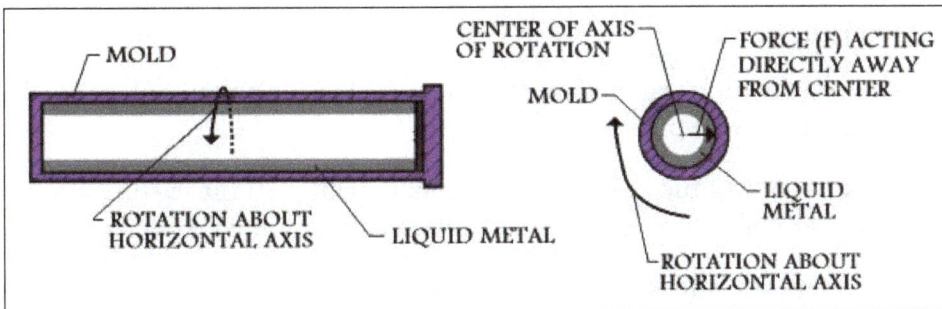

Solidification in true centrifugal casting.

It can be seen that this casting process is very well suited for the manufacture of hollow cylindrical tubes. The forces used in this technique guarantee good adhesion of the casting material to the surface of the mold. Thickness of the cast part can be determined by the amount of material poured. The outer surface does not need to be round. Polygonal geometries such as squares and other shapes can be cast. However, due to the nature of the process, the inner surface of a part manufactured by true centrifugal casting must always be round.

During the pouring and solidification phase of true centrifugal casting, the forces at work play a large roll in the properties of castings manufactured by this process. It can be seen that forces will be greater in the regions further away from the center of the axis of rotation. The greater forces towards the rim will cause the regions of the metal casting nearer the outer surface to have a higher density than the sections located nearer the inner surface.

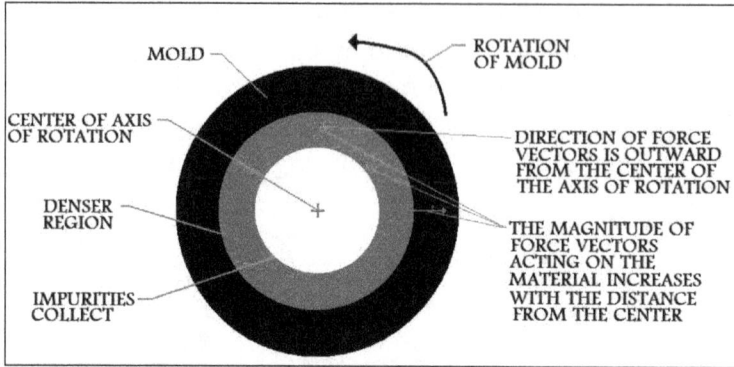

Force vector diagram for true centrifugal casting (cross-section).

Most impurities within the material have a lower density than the metal itself, this causes them to collect in the inner regions of the metal casting, closer to the center of the axis of rotation. These impurities can be removed during the casting operation or they can be machined off later.

Properties and considerations of manufacturing by true centrifugal casting:

- True centrifugal casting is a great manufacturing process for producing hollow cylindrical parts.

- The metal casting's wall thickness is controlled by the amount of material added during the pouring phase.

- Rotational rate of the mold during the manufacture of the casting must be calculated carefully based on the mold dimensions and the metal being cast.

- If the rotational rate of the mold is too slow, the molten material for the casting will not stay adhered to the surface of the cavity. From the top half of the rotation it will rain metal within the casting cavity as the mold spins.

- This manufacturing operation produces metal cast parts without the need for sprues, risers, or other gating system elements, making this a very efficient industrial metal casting process, in terms of material usage.

- Since large forces press the molten material for the cast part against the mold wall during the manufacturing operation, good surface finish and detail are characteristic of true centrifugal casting.

- Quality castings with good dimensional accuracy can be produced with this process.

- Material of high density and with few impurities is produced in the outer regions of cylindrical parts manufactured by true centrifugal casting.

- Impurities, such as metal inclusions and trapped air, collect in the lower density inner regions of cylindrical parts cast by this process.

- These inner regions can be machined out of the cast part leaving only the dense, more pure material.

- Shrinkage is not a problem when manufacturing by true centrifugal casting, since material from the inner sections will constantly be forced to instantly fill any vacancies that may occur in outer sections during solidification.

- This method can produce very large metal castings. Cylindrical pipes 10 feet in diameter and 50 feet long have been manufactured using this technique.

- With the employment of a sand lining in the mold, it is possible to manufacture castings from high melting point materials such as iron and steels.

- This is a large batch production operation.

- True centrifugal casting is a manufacturing process that is capable of very high rates of productivity.

Semicentrifugal Casting

Semicentrifugal casting manufacture is a variation of true centrifugal casting. The main difference is that in semicentrifugal casting the mold is filled completely with molten metal, which is supplied to the casting through a central sprue. Castings manufactured by this process will possess rotational symmetry. Much of the details of the manufacturing process of semicentrifugal casting are the same as those of true centrifugal casting. Parts manufactured in industry using this metal casting process include such things as pulleys, and wheels for tracked vehicles.

Process

In semi-centrifugal casting manufacture a permanent mold may be employed. However, often industrial manufacturing processes will utilize an expendable sand mold. This enables the casting of parts from high temperature materials.

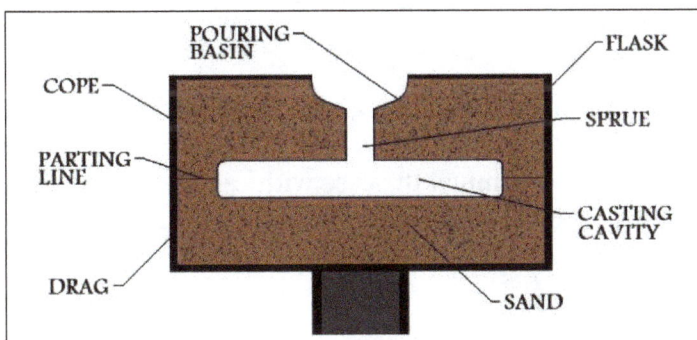

Semicentrifugal casting expendable sand mold used to manufacture a wheel.

The molten material for the metal casting is poured into a pouring basin and is distributed through a central sprue to the areas of the mold. The forces generated by the rotation of the mold ensure the distribution of molten material to all regions of the casting.

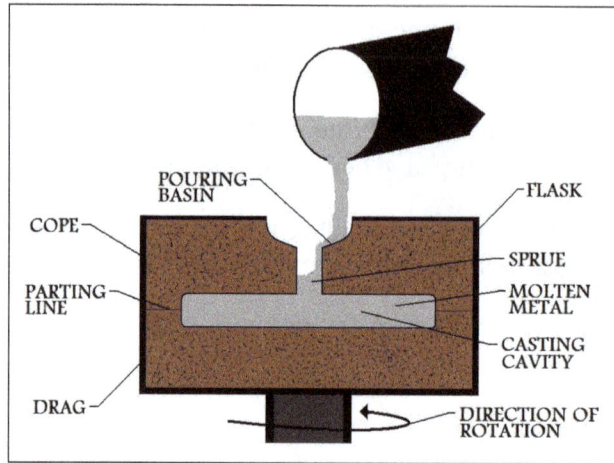

Semicentrifugal casting pouring of a wheel.

As the metal casting solidifies in a rotating mold, the centripetal forces constantly push material out from the central sprue/riser. This material acts to fill vacancies as they form, thus avoiding shrinkage areas.

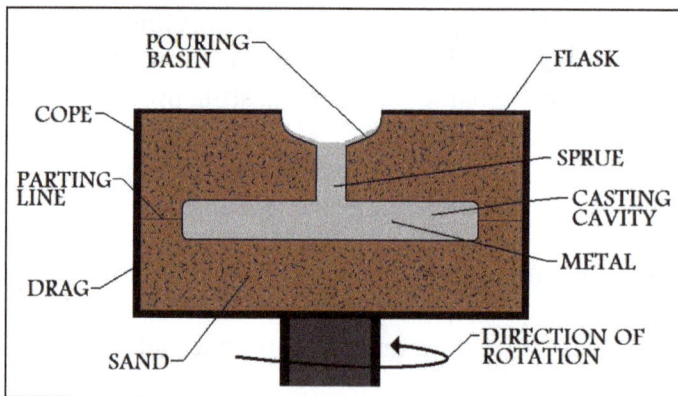

Semicentrifugal casting solidification of a wheel.

The centripetal forces acting on the casting's material during the manufacturing process of semicentrifugal casting, play a large part in determining the properties of the final cast part. This is also very much the case with cast parts manufactured using the true centrifugal casting process. Forces acting in the true centrifugal process are similar to those that influence the material of a metal casting being manufactured by semicentrifugal casting.

When manufacturing by semi-centrifugal casting, the centripetal acceleration generated on the mass of molten metal by a rotating mold is the force that acts to fill the casting with this molten metal. This is also the force that continues to act on the material as the casting solidifies. The main thing to remember about centripetal forces is that the force will push in a direction that is directly away from the center of the axis of rotation.

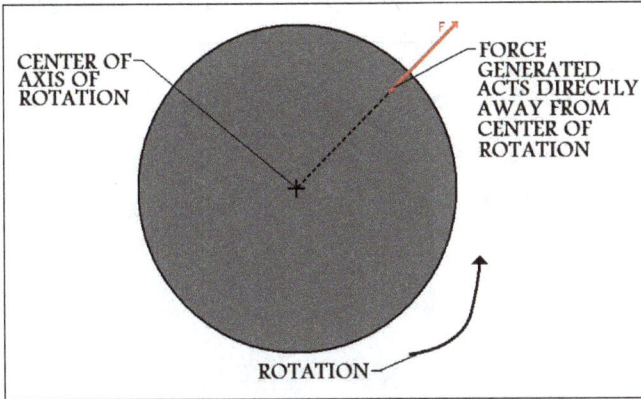

Spinning wheel.

Also, the farther away from the center of the axis of rotation, the greater the force.

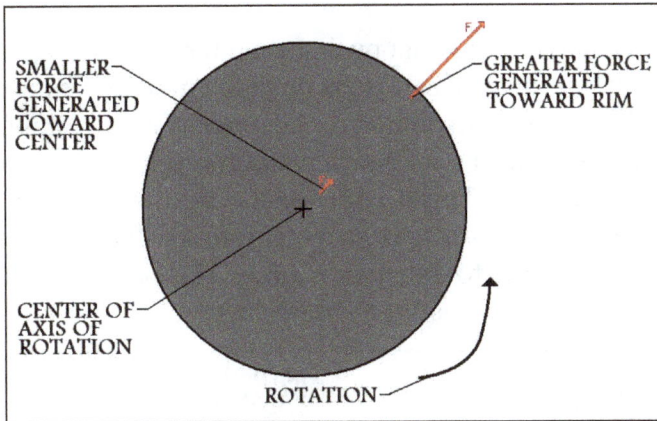

It can be seen that during the semicentrifugal manufacturing process, the material in the outer regions of the casting, (further from the center of the axis of rotation), is subject to greater forces than the material in the inner regions.

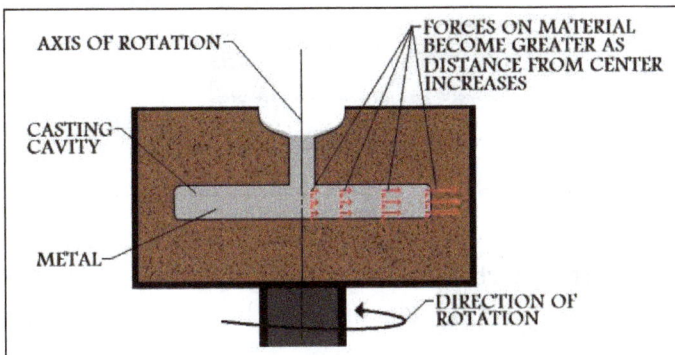

When the metal casting solidifies, the outer region of the cast part forms of dense material. The greater the forces under which the molten metal solidified, the denser the material in that region. So the density of a cast part manufactured by semicentrifugal casting will increase as you travel radially outward from the center.

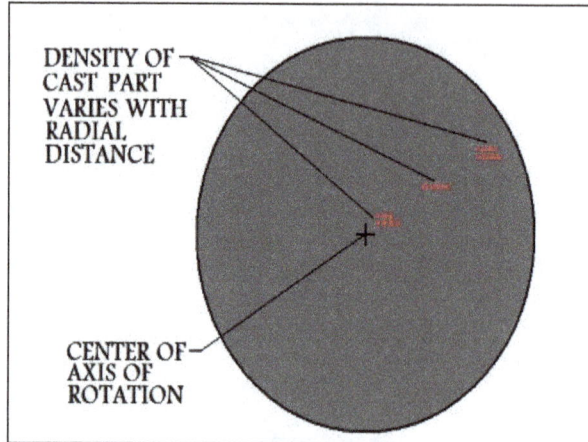

Cast wheel.

The high forces in the outer section that push the molten material against the mold wall also ensure a great surface finish of cast parts manufactured by semicentrifugal casting. Another feature of this process, attributed to the usage of centripetal forces, is that impurities within the metal, (such as solid inclusions and trapped air), will form towards the inner regions of the casting. This occurs because the metal itself is denser than the impurities, denser material subject to centripetal forces will tend to move towards the rim, forcing less dense material to the inner regions. This particular detail is also a feature in other types of centrifugal casting manufacture.

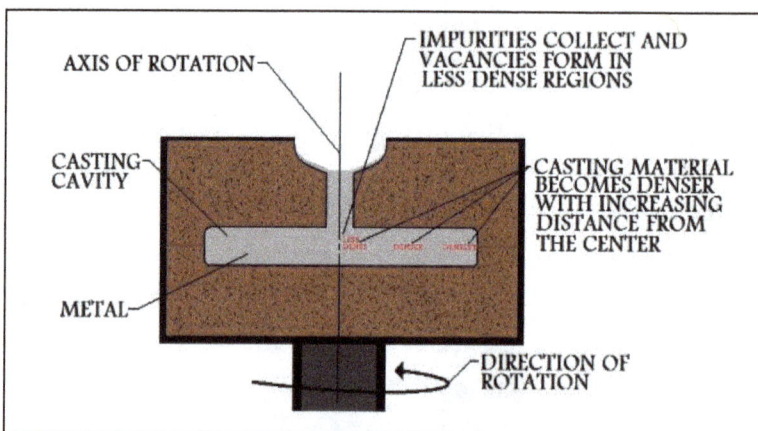

Semicentrifugal casting.

In industrial manufacture of parts by semicentrifugal casting, it is common to machine out the impurity filled center section, leaving only the purer, denser outer region as the final cast part.

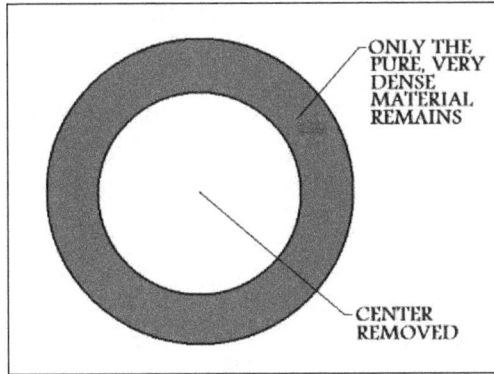

ONLY THE PURE, VERY DENSE MATERIAL REMAINS

CENTER REMOVED

Permanent Mold Casting

Permanent mold casting is the process in which the molds are made of metals like cast iron or steel. This casting technique is favored because mass production of castings can be done in a single run, thus helping in reducing manufacturing expenditure. In addition, permanent mold casting, which is an inexpendable casting process, can be fully automated.

In permanent mold casting molds are made of metals but the patterns can be made either of sand or metals. The molds are made using metals like cast iron or steel. These metal molds expand on heating or when hot metal is poured into it. Therefore, while making the mold, it is not expanded as is done with sand and plaster molds. But, care must be taken to maintain the thermal balance during the casting process.

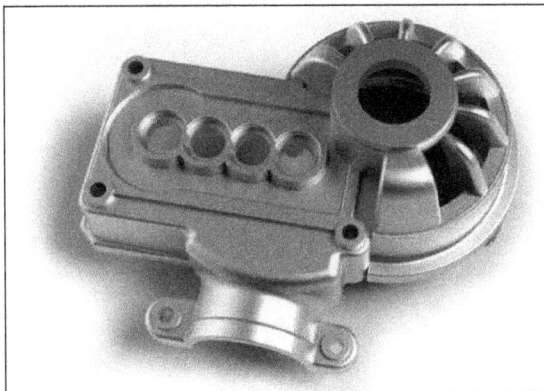

Metals are poured over the patterns to shape them into the required mold shapes. Unlike sand and plaster molds, permanent molds are not flexible to all types of patterns. Once the mold is set, the pattern is drawn out to reveal the coarse mold cavity. These cavities are to be machined to be made smooth. Cavity is then coated with a layer of refractory materials like clay or sodium silicate that makes the mold cavity heat resistant, allows easy ejection of the casting and increases the life of the mold. Machined gates are then attached to the mold.

For casting, the mold is first preheated and molten metal poured into the cavity and solidified. Once the metal is set, it is removed from the mold and the permanent mold is closed again to repeat the casting process. The casts take a weeks' time to solidify.

Types of Permanent Molds

- Gravity Permanent Mold: It the flow of the metal into the mold using the force of gravity. Gravity pouring are of two types: static pouring where the molten metal is poured form the top and; tilted pouring where the mold is slanted and the metal is poured into the mold using a basin. Gravity permanent mold casting produces accurate casting than shell mold castings.

- Low-Pressure permanent Mold: In this process only very little amount of forces is used to push the metal into the mold. Low-pressure permanent mold casting process enables producing uniform castings with excellent dimensional accuracy, perfect surface finish and superior mechanical properties.

Applications

Permanent mold casting process is used to cast products from iron, aluminum, magnesium, and copper based alloys. Typical permanent mold casting components include gears, splines, wheels, gear housings, pipe fittings, fuel injection housings, and automotive engine pistons, timing gears, impellers, compressors, pump parts, marine hardware, valve bodies, aircraft parts and missile components.

Advantages

- Suitable for high volume casting ceramic.

- Quality of heavier casting improves with better use of tooling's and equipment.

- Casted products have better tensile strength and elongation than sand castings.

- Mass productions can be done is a single production run, which reduces the manufacturing cost.

- Products have excellent mechanical properties.

Vacuum Permanent Mold Casting

Vacuum permanent mold casting is a permanent mold casting process employed in manufacturing industry that uses the force caused by an applied vacuum pressure to draw molten metal into and through the mold's gating system and casting cavity. This process has a similar name to vacuum mold casting however these are two completely different manufacturing processes and should not be confused with each other.

Process

A permanent mold containing the part geometry and the gating system is created, (usually accurately machined), similar to the molds employed in the other permanent mold processes. The mold in vacuum mold casting is much like the mold in the pressure casting manufacturing process, in that the gating system is designed so that the flow of molten material starts at the bottom and flows upwards.

Permanent mold for vaccum casting.

The mold is suspended over a supply of liquid metal for the casting by some mechanical device, possibly a robot arm.

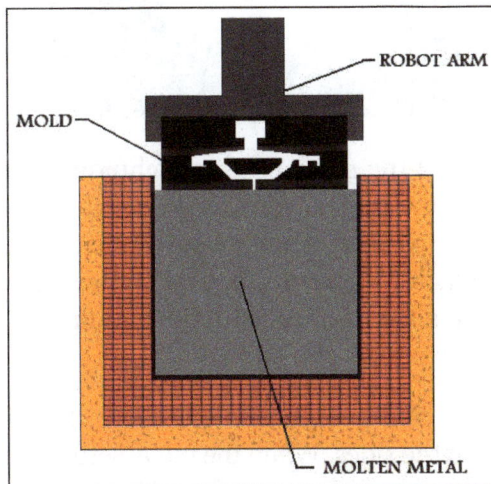

A vacuum force is applied to the top of the mold. The reduced pressure within the mold causes molten metal to be drawn up through the gating system and casting cavity.

As the casting solidifies, the mold is withdrawn from its position over the molten metal and opened to release the casting.

Properties and considerations of manufacturing by vacuum casting:

- This manufacturing process can produce metal castings with close dimensional accuracy, good surface finish, and superior mechanical properties.

- Castings with thin walled sections may be manufactured using this technique.

- This process is very much like pressure casting in the way the mold is filled, but since vacuum force is used instead of air pressure, gas related defects are reduced.

- Set up cost make this manufacturing process more suitable to high volume production, instead of small batch manufacture.

Slush Casting

Slush Casting is a traditional method of permanent mold casting process, where the molten metal is not allowed to completely solidify in the mold. When the desired thickness in obtained, the remaining molten metal in poured out. Slush casting method is an effective technique to cast hollow items like decorative pieces, components, ornaments, etc.

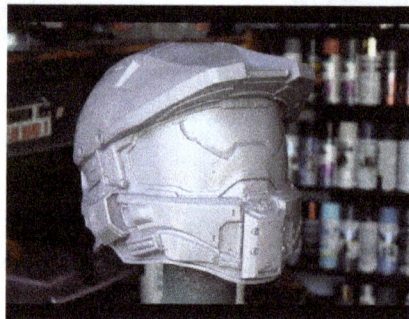

Mostly pewter is casted using the slush casting technique. Firstly, a pattern is made using plaster or wood. Now the pattern is placed on a cardboard or wooden board. A mold box is kept around the pattern. The unwanted space that is formed is the mold box can be eliminated by placing a board. Once the pattern is set the molding material is poured on the pattern and allowed to set with the molding aggregate. When the mold is set, the pattern is withdrawn from the mold.

The metal melted completely and poured into the mold which is shaped in the desired form. Rotate the mold to coat the sides. When the metal settles in the mold, remaining liquid metal is poured out of the mold. Thus, a hollow skin metal is formed inside the mold.

If the cast needs to be thicker, once again molten metal is poured into the mold and poured out. This process is repeated until the desired thickness is achieved. In some slush castings, bronze molds are used. When the metal hardens, the mold is broken to remove the castings. The inside of each cast retains molten textures while the exterior is smooth and shiny. Bowls and vases are serially produced by this technique that ensures no two are ever the same.

Similarly, to cast metals a bowl, a new process designed to capture the beauty of Pewter and its unique characteristics. Recycled molten Pewter is swirled inside amould to form a fine skin. The inside of each cast retains molten textures whilst the exterior is smooth and shiny. Bowls are serially produced by a technique that ensures no two are ever the same.

Application

Some casting of pewter is cast using slush casting method. Using pewter and other metal s mainly hollow products are casted. Decorative and ornamental objects that are casted are as vase, bowls, candlesticks, lamps, statues, jewelleries, animal miniatures, various collectibles, etc. Small objects and components for industry like tankard handle, handles for hollow wares, etc.

Advantage

- Slush casting is used to produce hollow parts without the use of cores.

- The desired thickness can be achieved by pouring our the left over molten metal.

- A variety of exquisitely designed casting can be casted for decorative and ornamental purpose.

Vacuum Molding

V-process or vacuum molding which was developed by Japanese using unbonded sand and vacuum is a perfect substitute for permanent mold and die casting process. Now the process is employed worldwide as an effective method to cast quality products in start up and low to medium job. The most highlighted feature of vacuum molding is that the flow of molten metal can be controlled.

Patterns are mounted on plates and boards, which are perforated, and each board is connected with a vacuum chamber. Unbonded sand is used for the molding purpose. Permeability is not a concern in this casting process, therefore sand of the finest structure can be used. The vented, plated pattern is coated with a layer of flexible plastic, which expand when the vacuum is applied in the mold. Enabling, the pattern to be stripped easily from the mold. Patterns should be perfectly smooth since in vacuum molding, every small intricate design gets imprinted on the cast. The pattern is not damaged during the process so they can be uses repeatedly.

In the vacuum molding process the mold are made is two parts (cope & drag) with each parts attached with its vacuum chamber. The pattern is kept and a metal or wooden flask place around it. Unbonded sand is poured over the molding box, and the tables are shaken vibrantly, by which the sand particle become tight and compact. Another layer of plastic sheet is draped over the molding box. The two halves are joined. Now the vacuum is formed through the patter. The vacuum makes the sand strong and the pattern coating expands, which makes it easy to strip the pattern from the mold.

The mold in kept in a housing and placed above a furnace of molten metal. Using sprue or gating the mold is connected inside the molten metal. When the vacuum from the mold is evacuated the molten metal gets forces into the mold, because of the difference in pressure that is created between the outer atmosphere and the mold. The plastic sheet melts and the mold is filled with the molten metal. After the metal solidifies and cools, the vacuum is released. The sand mold starts to fall apart as the solidification process completes. This sand can be cooled and reused for further casting process.

In mid 1600, Otto von Guericke a German mayor and scientist conducted the first experiment to prove the power of vacuum. He joined two large copper hemispheres and evacuated the air out of it. Now, eight horses were hooked on opposite side of the hemispheres. The horse pulled the hemisphere is two different direction, but the ball could not be torn apart. Guericke then let in air and the hemisphere came apart. In this way he proved the power and possibilities of vacuum.

Application

Vacuum molding process can be used to cast industrial components from both ferrous and non-ferrous metals.

Advantage

- Casted products have high dimensional accuracy and surface finish.

- The process is economical, environment friendly and clean.

- No moisture related defects for the castings.

- Provides consistent thickness for wall that give the casting an aesthetic appeal.

- Low cost operations.

Sand Casting

Sand casting, the most widely used casting process, utilizes expendable sand molds to form complex metal parts that can be made of nearly any alloy. Because the sand mold must be destroyed in order to remove the part, called the casting, sand casting typically has a low production rate. The sand casting process involves the use of a furnace, metal, pattern, and sand mold. The metal is melted in the furnace and then ladled and poured into the cavity of the sand mold, which is formed by the pattern. The sand mold separates along a parting line and the solidified casting can be removed.

Sand casting is used to produce a wide variety of metal components with complex geometries. These parts can vary greatly in size and weight, ranging from a couple ounces to several tons. Some smaller sand cast parts include components as gears, pulleys, crankshafts, connecting rods, and propellers. Larger applications include housings for large equipment and heavy machine bases. Sand casting is also common in producing automobile components, such as engine blocks, engine manifolds, cylinder heads, and transmission cases.

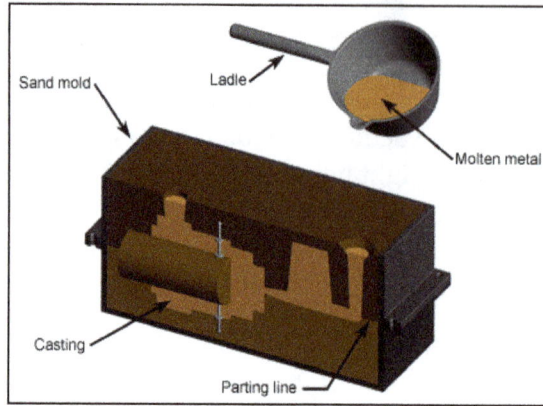

Process Cycle

The process cycle for sand casting consists of six main stages, which are explained below:

- Mold-making: The first step in the sand casting process is to create the mold for the casting. In an expendable mold process, this step must be performed for each casting. A sand mold is formed by packing sand into each half of the mold. The sand is packed around the pattern, which is a replica of the external shape of the casting. When the pattern is removed, the cavity that will form the casting remains. Any internal features of the casting that cannot be formed by the pattern are formed by separate cores which are made of sand prior to the formation of the mold. The mold-making time includes positioning the pattern, packing the sand, and removing the pattern. The mold-making time is affected by the size of the part, the number of cores, and the type of sand mold. If the mold type requires heating or baking time, the mold-making time is substantially increased. Also, lubrication is often applied to the surfaces of the mold cavity in order to facilitate removal of the casting. The use of a lubricant also improves the flow the metal and can improve the surface finish of the casting. The lubricant that is used is chosen based upon the sand and molten metal temperature.

- Clamping: Once the mold has been made, it must be prepared for the molten metal to be poured. The surface of the mold cavity is first lubricated to facilitate the removal of the casting. Then, the cores are positioned and the mold halves are closed and securely clamped together. It is essential that the mold halves remain securely closed to prevent the loss of any material.

- Pouring: The molten metal is maintained at a set temperature in a furnace. After the mold has been clamped, the molten metal can be ladled from its holding container in the furnace and poured into the mold. The pouring can be performed manually or by an automated machine. Enough molten metal must be poured to fill the entire cavity and all channels in the mold. The filling time is very short in order to prevent early solidification of any one part of the metal.

- Cooling: The molten metal that is poured into the mold will begin to cool and solidify once it enters the cavity. When the entire cavity is filled and the molten metal solidifies, the final shape of the casting is formed. The mold cannot be opened until the cooling time has elapsed. The desired cooling time can be estimated based upon the wall thickness of the casting and the temperature of the metal. Most of the possible defects that can occur are a result of the solidification process. If some of the molten metal cools too quickly, the part may exhibit shrinkage, cracks, or incomplete sections. Preventative measures can be taken in designing both the part and the mold.

- Removal: After the predetermined solidification time has passed, the sand mold can simply be broken, and the casting removed. This step, sometimes called shakeout, is typically performed by a vibrating machine that shakes the sand and casting out of the flask. Once removed, the casting will likely have some sand and oxide layers adhered to the surface. Shot blasting is sometimes used to remove any remaining sand, especially from internal surfaces, and reduce the surface roughness.

- Trimming: During cooling, the material from the channels in the mold solidifies attached to the part. This excess material must be trimmed from the casting either manually via cutting or sawing, or using a trimming press. The time required to trim the excess material can be estimated from the size of the casting's envelope. A larger casting will require a longer trimming time. The scrap material that results from this trimming is either discarded or reused in the sand casting process. However, the scrap material may need to be reconditioned to the proper chemical composition before it can be combined with non-recycled metal and reused.

Equipment

Mold

In sand casting, the primary piece of equipment is the mold, which contains several components. The mold is divided into two halves - the cope (upper half) and the drag (bottom half), which meet along a parting line. Both mold halves are contained inside a box, called a flask, which itself is divided along this parting line. The mold cavity is formed by packing sand around the pattern in each half of the flask. The sand can be packed by hand, but machines that use pressure or impact ensure even packing of the sand and require far less time, thus increasing the production rate. After the sand has been packed and the pattern is removed, a cavity will remain that forms the external shape of the casting. Some internal surfaces of the casting may be formed by cores.

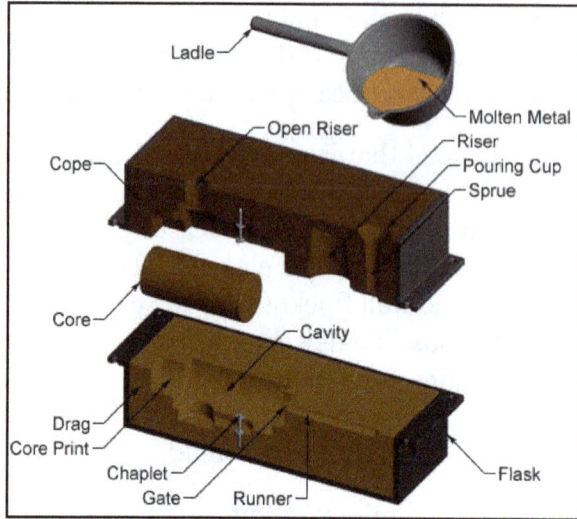
Sand Mold – Opened.

Cores are additional pieces that form the internal holes and passages of the casting. Cores are typically made out of sand so that they can be shaken out of the casting, rather than require the necessary geometry to slide out. As a result, sand cores allow for the fabrication of many complex internal features. Each core is positioned in the mold before the molten metal is poured. In order to keep each core in place, the pattern has recesses called core prints where the core can be anchored in place. However, the core may still shift due to buoyancy in the molten metal. Further support is provided to the cores by chaplets. These are small metal pieces that are fastened between the core and the cavity surface. Chaplets must be made of a metal with a higher melting temperature than that of the metal being cast in order to maintain their structure. After solidification, the chaplets will have been cast inside the casting and the excess material of the chaplets that protrudes must be cut off.

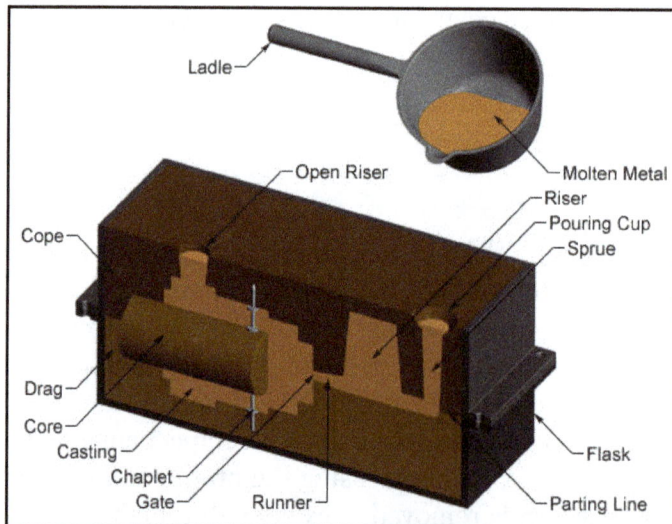
Sand Mold – Closed.

In addition to the external and internal features of the casting, other features must be incorporated into the mold to accommodate the flow of molten metal. The molten metal is poured into a pouring basin, which is a large depression in the top of the sand mold. The molten metal funnels out of the bottom of this basin and down the main channel, called the sprue. The sprue then connects to a series of channels, called runners, which carries the molten metal into the cavity. At the end of each runner, the molten metal enters the cavity through a gate which controls the flow rate and minimizes turbulence. Often connected to the runner system are risers. Risers are chambers that fill with molten metal, providing an additional source of metal during solidification. When the casting cools, the molten metal will shrink and additional material is needed. A similar feature that aids in reducing shrinkage is an open riser. The first material to enter the cavity is allowed to pass completely through and enter the open riser. This strategy prevents early solidification of the molten metal and provides a source of material to compensate for shrinkage. Lastly, small channels are included that run from the cavity to the exterior of the mold. These channels act as venting holes to allow gases to escape the cavity. The porosity of the sand also allows air to escape, but additional vents are sometimes needed. The molten metal that flows through all of the channels (sprue, runners, and risers) will solidify attached to the casting and must be separated from the part after it is removed.

Sand

The sand that is used to create the molds is typically silica sand (SiO_2) that is mixed with a type of binder to help maintain the shape of the mold cavity. Using sand as the mold material offers several benefits to the casting process. Sand is very inexpensive and is resistant to high temperatures, allowing many metals to be cast that have high melting temperatures. There are different preparations of the sand for the mold, which characterize the following four unique types of sand molds.

- Greensand mold - Greensand molds use a mixture of sand, water, and a clay or binder. Typical composition of the mixture is 90% sand, 3% water, and 7% clay or binder. Greensand molds are the least expensive and most widely used.

- Skin-dried mold - A skin-dried mold begins like a greensand mold, but additional bonding materials are added and the cavity surface is dried by a torch or heating lamp to increase mold strength. Doing so also improves the dimensional accuracy and surface finish, but will lower the collapsibility. Dry skin molds are more expensive and require more time, thus lowering the production rate.

- Dry sand mold - In a dry sand mold, sometimes called a cold box mold, the sand is mixed only with an organic binder. The mold is strengthened by baking it in an oven. The resulting mold has high dimensional accuracy, but is expensive and results in a lower production rate.

- No-bake mold - The sand in a no-bake mold is mixed with a liquid resin and hardens at room temperature.

The quality of the sand that is used also greatly affects the quality of the casting and is usually described by the following five measures:

- Strength - Ability of the sand to maintain its shape.

- Permeability - Ability to allow venting of trapped gases through the sand. A higher permeability can reduce the porosity of the mold, but a lower permeability can result in a better surface finish. Permeability is determined by the size and shape of the sand grains.

- Thermal stability - Ability to resist damage, such as cracking, from the heat of the molten metal.

- Collapsibility - Ability of the sand to collapse, or more accurately compress, during solidification of the casting. If the sand cannot compress, then the casting will not be able to shrink freely in the mold and can result in cracking.

- Reusability - Ability of the sand to be reused for future sand molds.

Packing Equipment

There exists many ways to pack the sand into the mold. As mentioned above, the sand can be hand packed into the mold. However, there are several types of equipment that provide more effective and efficient packing of the sand. One such machine is called a sand-slinger and fills the flask with sand by propelling it under high pressure. A jolt-squeeze machine is a common piece of equipment which rapidly jolts the flask to distribute the sand and then uses hydraulic pressure to compact it in the flask. Another method, called impact molding, uses a controlled explosion to drive and compact the sand into the flask. In what can be considered an opposite approach, vacuum molding packs the sand by removing the air between the flask and a thin sheet of plastic that covers the pattern. The packing of the sand is also automated in a process known as flask-less molding. Despite the name of the process, a flask is still used. In conventional sand casting, a new flask is used for each mold. However, flask-less molding uses a single master flask in an automated process of creating sand molds. The flask moves along a conveyor and has sand blown against the pattern inside. This automated process greatly increases the production rate and also has many benefits to the castings. Flask-less molding can produce uniform, high density molds that result in excellent casting quality. Also, the automated process causes little variation between castings.

Tooling

The main tooling for sand casting is the pattern that is used to create the mold cavity. The pattern is a full size model of the part that makes an impression in the sand mold.

However, some internal surfaces may not be included in the pattern, as they will be created by separate cores. The pattern is actually made to be slightly larger than the part because the casting will shrink inside the mold cavity. Also, several identical patterns may be used to create multiple impressions in the sand mold, thus creating multiple cavities that will produce as many parts in one casting.

Several different materials can be used to fabricate a pattern, including wood, plastic, and metal. Wood is very common because it is easy to shape and is inexpensive, however it can warp and deform easily. Wood also will wear quicker from the sand. Metal, on the other hand, is more expensive, but will last longer and has higher tolerances. The pattern can be reused to create the cavity for many molds of the same part. Therefore, a pattern that lasts longer will reduce tooling costs. A pattern for a part can be made many different ways, which are classified into the following four types:

- Solid pattern - A solid pattern is a model of the part as a single piece. It is the easiest to fabricate, but can cause some difficulties in making the mold. The parting line and runner system must be determined separately. Solid patterns are typically used for geometrically simple parts that are produced in low quantities.

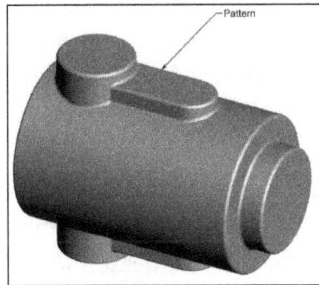

Solid pattern.

- Split pattern - A split pattern models the part as two separate pieces that meet along the parting line of the mold. Using two separate pieces allows the mold cavities in the cope and drag to be made separately and the parting line is already determined. Split patterns are typically used for parts that are geometrically complex and are produced in moderate quantities.

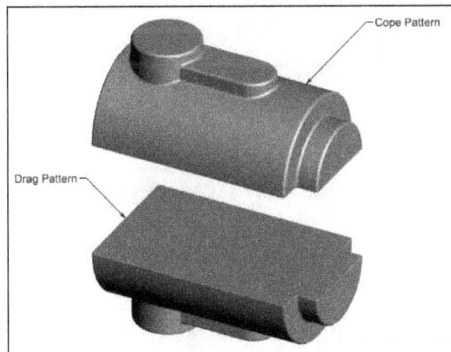

Split pattern.

- Match-plate pattern - A match-plate pattern is similar to a split pattern, except that each half of the pattern is attached to opposite sides of a single plate. The plate is usually made from wood or metal. This pattern design ensures proper alignment of the mold cavities in the cope and drag and the runner system can be included on the match plate. Match-plate patterns are used for larger production quantities and are often used when the process is automated.

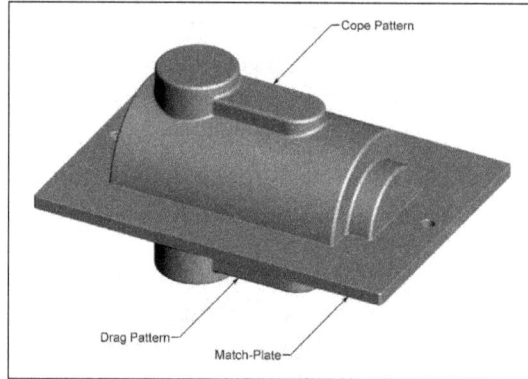

Match-plate pattern.

- Cope and drag pattern - A cope and drag pattern is similar to a match plate pattern, except that each half of the pattern is attached to a separate plate and the mold halves are made independently. Just as with a match plate pattern, the plates ensure proper alignment of the mold cavities in the cope and drag and the runner system can be included on the plates. Cope and drag patterns are often desirable for larger castings, where a match-plate pattern would be too heavy and cumbersome. They are also used for larger production quantities and are often used when the process is automated.

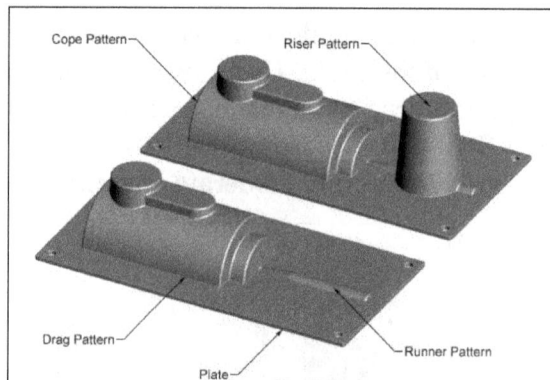

Cope and drag pattern.

Another piece of tooling used in sand casting is a core-box. If the casting requires sand cores, the cores are formed in these boxes, which are similar to a die and can be made of wood, plastic, or metal just like the pattern. The core-boxes can also contain multiple cavities to produce several identical cores.

Materials

Sand casting is able to make use of almost any alloy. An advantage of sand casting is the ability to cast materials with high melting temperatures, including steel, nickel, and titanium. The four most common materials that are used in sand casting are shown below, along with their melting temperatures.

Materials	Melting temperature
Aluminum alloys	1220 °F (660 °C)
Brass alloys	1980 °F (1082 °C)
Cast iron	1990-2300 °F (1088-1260 °C)
Cast steel	2500 °F (1371 °C)

Possible Defects Causes

Unfilled Sections

- Insufficient material.

- Low pouring temperature.

Porosity

- Melt temperature is too high.

- Non-uniform cooling rate.

- Sand has low permeability.

Hot Tearing

- Non-uniform cooling rate.

Surface Projections

- Erosion of sand mold interior.

- A crack in the sand mold.

- Mold halves shift.

Design Rules

Maximum Wall Thickness

- Decrease the maximum wall thickness of a part to shorten the cycle time (cooling time specifically) and reduce the part volume.

Incorrect

Correct

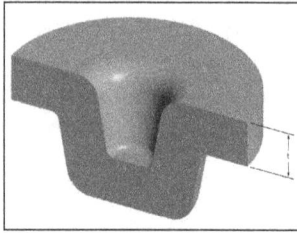

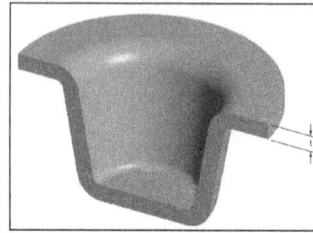

Part with thick walls.

Part redesigned with thin walls.

- Uniform wall thickness will ensure uniform cooling and reduce defects. A thick section, often referred to as a hot spot, causes uneven cooling and can result in shrinkage, porosity, or cracking.

Incorrect

Correct

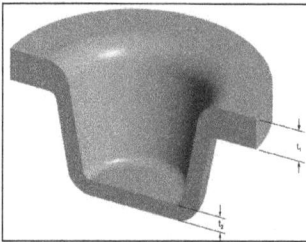

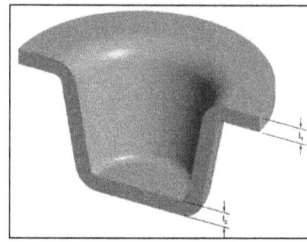

Non-uniform wall thickness ($t_1 \neq t_2$).

Uniform wall thickness ($t_1 = t_2$).

Corners

- Round corners to reduce stress concentrations and fracture.

- Inner radius should be at least the thickness of the walls.

Correct

Incorrect

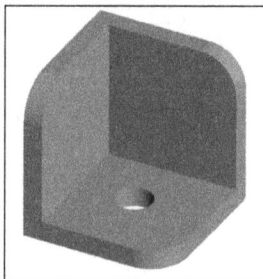

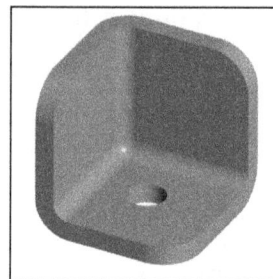

Sharp corner.

Rounded corner.

Draft

- Apply a draft angle of 2° - 3° to all walls parallel to the parting direction to facilitate removing the part from the mold.

Incorrect   Correct

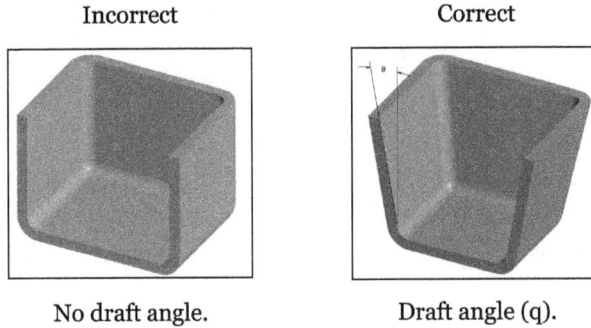

No draft angle.   Draft angle (q).

Machining Allowance

- Add 0.0625 - 0.25 in. (0.16 - 0.64 mm) to part dimensions to allow for machining to obtain a smooth surface.

Cost Drivers

Material Cost

The material cost for sand casting includes the cost of the metal, melting the metal, the mold sand, and the core sand. The cost of the metal is determined by the weight of the part, calculated from part volume and material density, as well the unit price of the material. The melting cost will also be greater for a larger part weight and is influenced by the material, as some materials are more costly to melt. However, the melting cost in typically insignificant compared to the metal cost. The amount of mold sand that is used, and hence the cost, is also proportional to the weight of the part. Lastly, the cost of the core sand is determined by the quantity and size of the cores used to cast the part.

Production Cost

The production cost includes a variety of operations used to cast the part, including core-making, mold-making, pouring, and cleaning. The cost of making the cores depends on the volume of the cores and the quantity used to cast the part. The cost of the mold-making is not greatly influenced by the part geometry when automated equipment is being used. However, the inclusion of cores will slightly slow the process and therefore increase the cost. Lastly, the cost of pouring the metal and cleaning the final casting are both driven by the weight of the part. It will take longer to pour and to clean a larger and heavier casting.

Tooling Cost

The tooling cost has two main components - the pattern and the core-boxes. The pattern cost is primarily controlled by the size of the part (both the envelope and the projected area) as well as the part's complexity. The cost of the core-boxes first depends on their size, a result of the quantity and size of the cores that are used to cast the part. Much like the pattern, the complexity of the cores will affect the time to manufacture this part of the tooling (in addition to the core size), and hence the cost.

The quantity of parts that are cast will also impact the tooling cost. A larger production quantity will require the use of a tooling material, for both the pattern and core-boxes, that will not wear under the required number of cycles. The use or a stronger, more durable, tooling material will significantly increase the cost.

Sand Moulds

A sand moulds may be defined as-a preformed sand container into which molten metal is poured and allowed to solidify. After casting it is removed from the sand mould, sand mould is generally destroyed. The moulds is filled by pouring the molten metal into an opening at the top of the mould and proper passages are made to allow the metal to flow to all the parts of the mould by gravity.

Small or medium sized castings are generally made in a flask—a rectangular box-shaped container, without top and bottom. The flask may be made in two or three parts, and parts are held in alignment by dowel pins. It is necessary to clamp the flask before pouring molten metal into it, in order to prevent the buoyant effect of the molten metal from lifting up the top part of flask.

Classification of Sand Moulds

Depending upon the material used, the moulds could be classified as:

- Green sand moulds,
- Skin-Dry moulds,
- Dry sand moulds,
- Cement-Bonded moulds,
- Metal moulds.

Green Sand Moulds

Green sand moulds are those sand moulds, in which moisture is present in the sand at the time of pouring the molten metal. The grains are held together by moist clay. Moisture level has to be controlled carefully. These are used for casting, practically all ferrous alloys. Green sand is available in many kinds and is used for making small, medium and often large moulds even.

Green sand moulds are least expensive to make as the basic material for these is cheaper. Larger output can be obtained from given floor space. These do not require any backing operations or equipment but dry sand cores are to be used. These being softer than dry sand moulds, allow greater freedom in contraction, when the castings solidify and cool.

Also, the moulding is less time-consuming. However, green sand moulds have some disadvantages viz., they not being as strong as others are liable to be damaged during handling or by metal erosion. The moisture present in the sand may also cause certain defects in the casting like blow holes, gas holes etc.

These moulds cannot be stored for long time. The surface finish of the casting obtained from green sand mould is not very smooth. Sometimes additives like coal dust or organic materials are also added and then it is called loam moulding.

The three commonly used methods of green-sand moulding are:

Open Sand Method

It is simplest form in which, the entire mould is made in the foundry floor or in a bed of sand above floor level. This method is mainly employed for simple solid castings with flat tops.

After proper levelling, the pattern is pressed in the sand bed for making mould. Moulding box is not necessary and the upper surface of the mould is open to air. Pouring basin is made at one end of the mould, and the overflow channel cut at the sides of the cavity.

Bedded-in Method

In this method a cope i.e. a sand cover is necessary. It is used, when the upper surface of casting is not flat. The pattern is hammered down the sand of the foundry floor or in a drag filled partially with sand to form the mould cavity. The top of drag is smoothened and the parting sand spreaded. A cope is placed over the pattern and rammed up.

Runners and risers are cut and the cope box is lifted. The pattern is then withdrawn, the surfaces of drag and cope moulds finished and cope replaced in its correct position for completing the mould.

Turn-over Method

This method is commonly used for solid as well as split patterns. One half of the pattern is placed with its flat side on a moulding board, a drag is rammed and rolled over. Next, the cope is placed over the other half of the pattern and is rammed and rolled over. The two pattern halves are shaken and withdrawn. Now, the cope is placed on the drag for assembling the mould.

Skin-dry Moulds

These are made of green sand with dry sand baking. In some cases, moisture is dried from the surface layer of rammed sand to a depth of 25 mm by heater or gas torches. These are more common in large moulds and can be used for casting, practically all ferrous and non-ferrous alloys.

These are less expensive to construct than dry-sand moulds but more expensive than green sand moulds of given size. It has the advantages of less equipment, cheaper materials, less time for preparation, and less floor space in comparison to dry sand moulding. However, these are not as strong as dry sand moulds and can't be stored for long time as moisture may migrate through the dry skin.

Dry Sand Moulds

These are made with that sand which doe" not require moisture to develop strength. The sand mixture for small and medium works consists of 13 parts of floor sand, 8 parts of new sand and 1 part of horse manure or saw dust. For heavy work, these proportions are 11:9:1 and for extra heavy work—10:10:1.

The mould surface is sprayed with molasses water. All parts of mould are baked in furnace at 150—300 °C (until moisture is driven off) to increase the strength, resist erosion, and improve surface conditions. The dry-sand moulds may be used for many alloys but are more commonly used for steel castings.

These are used mostly in small and medium sized operations. For larger sized operations, the dry sand moulds are made in sections and assembled after baking. The dry sand moulds are stronger and can be handled more easily with less damage and also can be stored for longer time.

These resist metal erosion, and tendency for moisture related defects is eliminated. The disadvantages of these moulds are- these require more expensive moulding material, labour costs are high, and extra operation, equipment and space are needed.

Cement-bonded Moulds

In these moulds, silica sand bonded with Portland cement is used as the moulding material, which dries up in air. These moulds are most commonly used for very large ferrous work and pit moulding and in other cases, where baking is impossible. It has high strength and possesses all advantages of dry sand.

For these moulds, extra space for air drying operation has to be provided. The materials used in these moulds can't be used again like other moulds, thus the process becomes expensive.

Metal Moulds

These are used for die-casting, permanent mould and centrifugal casting processes.

Feeding of Metal in Sand Moulds

To take care of feeding problems, particularly when casting is complex in size and shape, is the most important aspect to ensure sound castings. Problems are encountered because of slow rate of heat abstraction from a large mass of metal. This results in minimising temperature gradients and difficulty in obtaining solidification in directional manner towards the feeder.

While, it is easy to obtain directional solidification with alloys which have short freezing range, it is quite difficult with alloys having wide freezing range. Location of feeders and risers, to supply hot liquid to thicker sections, which may become isolated during solidification, and to establish temperature gradients initially, is very important. Sometimes temperature gradients can be established artificially by using chills within the moulds, insulating pads and tapered sections.

It may be noted that shrinkage cavities are likely to form in sections like T_s and crosses etc. which take longer to solidify because in such cases, volume of metal is more and surface area through which heat can escape is reduced. The solution, therefore, lies in providing shapes in which volume of metal is decreased and the surface area through, which heat can escape is increased.

In the case of crosses, best results can be obtained either by staggering the location of ribs, or introducing a cored hole at the centre of cross, or using a circular web whose walls are thinner than straight ribs. It must be ensured that last portion to solidify will be fed with metal from feeder. Design, sizing and location of feeders, gates and channels is, therefore, most important.

Usually feeding problems are best solved by experience, yet some scientific and empirical rules developed with experience will be found to serve a good guide in tackling this problem.

Heat Transfer Approach

Chvorinov's Rule

For any shape where the same interface boundary conditions hold, according to Chvorinov's rule, solidification time is directly proportional to the square of the ratio of volume to surface area.

For same volume, solidification time will progressively increase for the following shapes in order because their surface areas for same volumes decrease:

- Plates,
- bars,
- cuboids,

- short cylinders,

- spheres.

Though, theoretically sphere takes longest to solidify but is not feasible for feeders. Short cylinder is nearest approached and it is tapered to facilitate moulding.

Evaporative Pattern Casting

Evaporative casting, consumable or eva-foam casting is a sand casting process where the foam pattern evaporates into the sand mold. A process similar to investment casting, this expendable casting process is predicted to be used for 29% of aluminum and 14% of ferrous casting in 2010. There are two main evaporative casting process lost-foam casting and full moldcasting which are widely used because intricate design can be cast with relative ease and with reasonable expense. The main difference between the two is that in the lost-foam casting unbonded sand is used and in the full-mold casting green sand or bonded sand is used.

In the first step of evaporative casting, a foam pattern is shaped using material like polystyrene. The pattern is attached with sprues, and gates using adhesives and brushed with refractory substances so that the molds are strong and resistant to high temperature. Refractory covered pattern assembly is then surrounded by a sand mixture to form a mold. In some instances the pattern assembly is mixed in ceramic slurry which forms a shell round the pattern when it dries.

In both cases, the mold in kept at a specific temperature to allow the metal to flow smoothly and enter into every designs and cuts made by the pattern. Molten metal is poured into the mold and the pattern-forming material disappears into the mold. The molten metal takes the shape of the mold and solidifies. When the metal solidifies it is removed from the mold to form the casting.

Unlike in the traditional sand casting method, in evaporative sand casting, the pattern does not have to be removed from the mold which reduces the need for draft provisions.

Some of the parameter that are used to determine the quality of a eva-foam casting are grain fineness number, time of vibration, degree of vacuum and pouring temperature on surface roughness etc.

Applications

Evaporative castings is used for steel-casting cast iron parts like water pipe and pump parts, aluminum castings etc.

Advantage

- High dimensional accuracy and superior casting surface smoothness.

- Reduced work process unlike other casting methods.

- Light weight casting are be done.

- Casting have improved heat resistance and also abrasion resistance and other cast steel properties.

- Complicated shapes can be cast without using cores or drafts.

Investment Casting

Investment casting, also known as precision casting or lost-wax casting, is a manufacturing process in which a wax pattern is used to shape a disposable ceramic mold. A wax pattern is made in the exact shape of the item to be cast. This pattern is coated with a refractory ceramic material. Once the ceramic material is hardened, it is turned upside-down and heated until the wax melts and drains out. The hardened ceramic shell becomes an expendable investment mold. Molten metal is poured into the mold and is left to cool. The metal casting is then broken from of the spent mold.

The term investment casting is derived from the process of "investing" (surrounding) a pattern with refractory materials. Investment casting is often selected over other molding methods because the resulting castings present fine detail and excellent as-cast surface finishes. They can also be cast with thin walls and complex internal passageways. Unlike sand casting, investment casting does not require a draft.

These process qualities can provide net shape or near-net shape castings, which provide customers with significant cost savings in material, labor, and machining. It can make use of most common metals, including aluminum, bronze, magnesium, carbon steel, and stainless steel. Parts manufactured with investment casting include turbine blades, medical equipment, firearm components, gears, jewelry, golf club heads, and many other machine components with complex geometry.

The "Looking Up" sculpture in New York uses investment casting
techniques to create an impactful stainless steel figure.

Investment Casting Process

The investment casting process consists of several steps: metal die construction, wax pattern production, ceramic mold creation, pouring, solidification, shakeout, and cleanup.

Metal Die Construction

The wax pattern and ceramic mold are destroyed during the investment casting process, so each casting requires a new wax pattern. Unless investment casting is being used to produce a very small volume (as is common for artistic work or original jewelry), a mold or die from which to manufacture the wax patterns is needed.

The size of the master die must be carefully calculated; it must take into consideration expected shrinkage of the wax pattern, the expected shrinkage of the ceramic material invested over the wax pattern, and the expected shrinkage of the metal casting itself.

Wax Pattern Production

The number of wax patterns always equals the number of castings to be produced; each individual casting requires a new wax pattern.

Hot wax is injected into the mold or die and allowed to solidify. Cores may be needed to form any internal features. The resulting wax pattern is an exact replica of the part to be produced. The method is similar to die-casting, but with wax used instead of molten metal.

Hot wax is injected into a mold or die and allowed to solidify, resulting
in a wax pattern that is an exact replica of the part to be produced.

Mold Creation

A gating system (sprue, runner bars, and risers) is attached to the wax mold. For small-er castings, several wax patterns are attached to a central wax gating system to form a tree-like assembly. A pouring cup, typically attached to the end of the runner bars, serves to introduce molten metal into the mold.

The assembled "pattern tree" is dipped into a slurry of fine-grained silica. It is dipped repeatedly, being coated with progressively more refractory slurry with each dip. Once the refractory coating reaches the desired thickness, it is allowed to dry and harden; the dried coating forms a ceramic shell around the patterns and gating system.

Once the coating of slurry reaches the desired thickness, it is left
to dry and harden, forming a ceramic shell around the pattern.

The thickness of the ceramic shell depends of the size and weight of the part being cast, and the pouring temperature of the metal being cast. The average wall thickness is approximately 0.375 in. (9.525 mm). The hardened ceramic mold is turned upside down, placed in an oven, and heated until the wax melts and drains away. The result is a hollow ceramic shell.

Pouring

The ceramic mold is heated to around 1000 – 2000°F (550 – 1100°C). The heating process further strengths the mold, eliminates any leftover wax or contaminants, and evaporates water from the mold material.

Molten metal is poured into the mold while it is still hot – liquid metal flows into the pouring cup, through the central gating system, and into each mold cavity on the tree. The pre-heated mold allows the metal to flow easily through thin, detailed sections. It also creates a casting with improved dimensional accuracy, as the mold and casting will cool and shrink together.

Cooling

After the mold has been poured, the metal cools and solidifies. The time it takes for a mold to cool into a solid state depends on the material that was cast and the thickness of the casting being made.

Shakeout

Once the casting solidifies, the ceramic molds break down, and the casting can be removed. The ceramic mold is typically broken up manually or by water jets. Once removed, the individual castings are separated from their gating system tree using manual impact, sawing, cutting, burning, or by cold breaking with liquid nitrogen.

Finishing

Finishing operations such as grinding or sandblasting are commonly employed to smooth the part at the gates and remove imperfections. Depending on the metal that the casting was poured from, heat treating may be employed harden the final part.

Sandblasting is commonly employed to clean and
smooth out metal parts to remove imperfections.

When to use Investment Casting

Due to its complexity and labor requirements, investment casting is a relatively expensive process – however the benefits often outweigh the cost. Practically any metal can be investment cast. Parts manufactured by investment casting are normally small, but the process can be used effectively for parts weighing 75 lbs or more.

Investment casting is capable of producing complex parts with excellent as-cast surface finishes. Investment castings do not need to have taper built in to remove the components from their molds because the ceramic shells break away from the part upon cooling. This production feature allows castings with 90-degree angles to be designed with no shrinkage allowance built-in, and with no additional machining required to obtain those angles.

The investment casting process creates parts with superior dimensional accuracy; net-shape parts are easily achievable, and finished forms are often produced without secondary machining. Each unique casting run requires a new die to produce wax patterns. Tooling for investment casting can be quite expensive; depending on the complexity, tooling costs can run anywhere between $1000 and $10,000.

For high volume orders, the time and labor saved by eliminating or decreasing secondary machining easily makes up for the cost of new tooling. Small casting runs are less likely to make up for the investment. Generally, investment casting is a logical choice for a run of 25 parts or more.

Investment casting is used to create parts with superior dimensional accuracy,
where finished forms are often produced without secondary machining.

It usually takes 7 days to go from a fresh wax pattern to a complete casting; the majority of that time is taken up by creating and drying the ceramic shell mold. Some foundries have quick-dry capabilities to produce castings more quickly. The time and labor-intensive nature of investment casting doesn't only effect cost. Foundries have limited equipment and production capacity, so longer lead times for investment casting are common.

Plastic Casting

Casting involves introducing a liquefied plastic into a mold and allowing it to solidify. In contrast to molding and extrusion, casting relies on atmospheric pressure to fill the mold rather than using significant force to push the plastic into the mold cavity. Some polymers have a viscosity similar to bread dough even when they are at elevated temperature so they are not candidates for the casting process. Examples of this are polymers like POM, PC, PP and many others. Casting includes a number of processes that take a monomer, powder or solvent solution and pur them into a mold. They transition from liquid to solid by either evaporation, chemical action, cooling or external heat. The final product can be removed from the mold once it solidifies.

Advantages

- Cost of equipment, tooling and molds are low.

- The process is not complex.

- Products have little or no internal stress.

Disadvantages

- The output rate is slow and has long cycle times.

- Dimensional tolerances are not very good.

- Moisture and air bubbles can be difficult to manage and may cause problems.

There is a vast array of materials than can be cast. Nylon Type 6 is one of the most popular and commonly used cast products.

Resin Casting

Resin casting or plastic resin casting is a form of plastic casting process in which the molding materials used are liquid resin like acrylics, polyesters, urethanes, epoxies, etc. This process is primarily used for casting industrial prototypes and dentistry. Plastic resin casting is an uncomplicated casting procedure that is preferred by hobbyist and artists.

Plastic resin casting can be done at normal room temperature and pressure using molds made from latex rubber, vulcanized silicone rubber or other materials that are quite cheap and affordable. Liquid resin which is mixed with a curing agent is poured into the mould at room temperature. The bubbles that are created during the molding process can be expelled by shaking the mold. Resin harden to take the intricate shape of the mold.

The synthetic resins used include polyurethane resin, epoxy resin, unsaturated polyester resin and silicone resin. A flexible mold can be made of latex rubber, room temperature vulcanized silicone rubber or other similar materials at relatively low cost. The mold are commonly made in two halves.

The simplest method is gravity casting where the resin is poured into the mold and pulled down into all the parts by gravity. Hardened resin casting is removed from the flexible mold and allowed to cool.

Casting can also be done in a vacuum chamber or in a pressure pot, to reduce their size to the point where they aren't visible. Pressure and centrifugal force can be used to help push the liquid resin into all details of the mold. The mold can also be vibrated to expel bubbles.

Applications of Resin Casting

Resin casting is an excellent media for reproducing any form for objects. Hence, this process is used to cast prototypes to complete model of ships, cars, trains, air crafts, buildings, etc. Resin can be cast is intricate design using a variety of colour combinations. Therefore, this medium is used for casting beautiful jeweleries, attractive toys, etc. Bangles and other ornamentals accessories using resin is quite popular among college goers.

Advantages of Resin Casting

- Plastic resin casting enables the casting of intricate designs.

- Can be casted or painted in any desired colour.

- Finished models of toys, jeweleries and protype castings is an exact replica of the original ones using costly materials.

Glass Casting

Glass casting is a simple but effective means of shaping glass items. First, you heat up the temperature of the glass to the point where it gets soft and malleable. Then, you transfer the liquid glass into a mold. This mold may be made of metal, graphite, or even sand.

The glass fills the mold and hardens. After the glass has totally cooled, remove the glass from the mold. You can then enjoy your glass item. It's as simple as that.

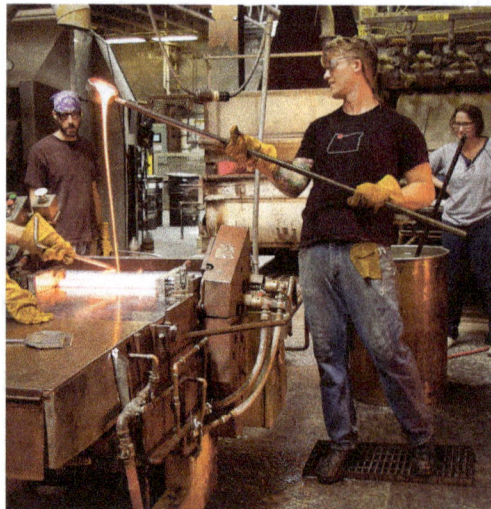

The process of glass casting has a long, fascinating history. This technique goes back to the Roman and Egyptian times. In its earliest iteration, the Romans would make glass dishes through glass casting. The technique caught on, moving next to Italy and later, the rest of the world.

Working of Glass Casting

Depending on the type of mold you're using, glass casting techniques differ somewhat. We'll explain each technique:

- Graphite casting: Graphite molds can be various shapes. Glass must be heated at very high temperatures until it's a molten liquid. It's then added to the molds. Before the glass solidifies, it must be moved to an annealing kin.

- Pate de verre: This French term refers to using glass paste to make shaped glass. Enamels, colorants, water, and gum arabic are combined with miniscule glass pieces. This creates a sticky texture that can be fitted inside a mold. This mold is malleable due to the texture of the pate de verre.

- Kiln casting: Silica or plaster molds are most commonly used with kiln casting. Glass isn't the only material that can go into a kiln mold; metal, wood, and sometimes wax are favored, too.

- Sand casting: Sand casting is similar to the process described above. The mold already comes prepared. The glass is heated until it's a liquid. It sticks to the sand mold due to the clay bentonite, which prevents absorption of liquids. Once the glass hardens, it can be removed from the mold.

References

- Casting-of-metals-top-11-methods-industries-metallurgy, casting, metallurgy: engineeringenotes. com, Retrieved 26 July, 2019

- Continuous-casting: the library of manufacturing.com, Retrieved 21 May, 2019

- Die-casting, wu: custompartnet.com, Retrieved 8 January, 2019

- Types-of-die-casting-machines-with-diagram, metal-castings, metallurgy: yourarticlelibrary.com, Retrieved 13 May, 2019

- Hot-chamber-diecasting: thelibraryofmanufacturing.com, Retrieved 16 January, 2019

- Pressure-die-casting: themetalcasting.com, Retrieved 25 February, 2019

- Gravity-casting-gravity-casting-process: ferralloy.com, Retrieved 29 March, 2019

- Casting-slush-casting: industrialmetalcastings.com, Retrieved 13 February, 2019

- Sand-moulds-definition-and-classification-casting-metallurgy, casting, metallurgy: engineeringe-notes.com, Retrieved 17 May, 2019

- Investment-casting: reliance-foundry.com, Retrieved 13 July, 2019

Casting Properties

There are various properties of different material which have to be considered during the casting process. A few of these are mechanical properties of cast iron, tensile properties of aluminum foundry alloys and the fracture toughness of metal castings. The topics elaborated in this chapter will help in gaining a better perspective about these casting properties.

Mechanical Properties of Heavy Section Castings

The effect of section size on properties in castings can be separated into either a geometrical or metallurgical size effect. The geometrical size effect is apparent when testing different size specimens with the same metallurgical origin. On the other hand, the metallurgical size effect is the testing of similarly sized specimens machined from castings of different sizes. In heavy section casting, both section size effects are evident. An understanding of both types of size effects will help the producer minimize the adverse impact of section size on the service life of the casting.

The properties of interest to the modern designer include tensile, impact, fracture toughness, and fatigue. Yield strength was the classical engineering property used as a basis for design. However, most components that fail, fail starting from a flaw and exhibit an absence of plastic deformation or yielding. The failure may have occurred starting at the flaw with a single load application (fracture toughness) or the flaw may have provided a site for a crack which grew to critical size only after multiple applications of the load (fatigue and fracture toughness). Fracture toughness and fatigue tests are successful in allowing design calculations to avoid these failures.

Metallurgical Size Effect

The metallurgical size effect is attributed to the changes of microstructure inherent in producing and heat treating different size castings. Included in this category are normal effects, like changes in grain size, and lack of through-hardening; and defects more prone to occur in large cast sections such as large inclusion size, temper embrittlement, rock candy, microshrinkage, surface roughness, and surface pick-up. All of these effects normally increase as the section size of the casting increases.

Grain Size and Heat Treating Effects

The grain size of steel increases with an increase in casting section size as given in table. In general, the mechanical properties of a steel are related to the grain size. Figure shows the benefits of grain refinement on the tensile properties of mild steel. As the grain size becomes smaller - the tensile strength increases. Similarly, figure, shows how the grain size affects the fracture strength, a fracture mechanic's measure of the resistance to crack propogation.

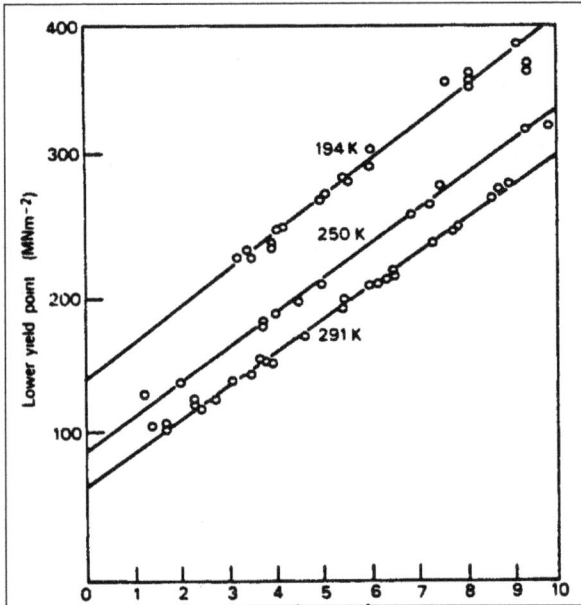

Dependence of the lower yield stress of mild steel of grain size.

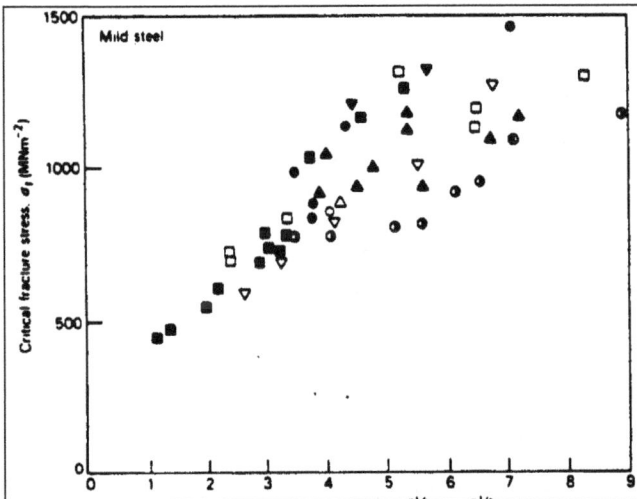

Dependence of local fracture stress of on the grain size of mild steel.

The fatigue behavior of steel is also affected by the grain size. The fatigue limits of two steels with different grain sizes are compared in figure. There was an increase of endurance limit with a decrease in grain size. Larger grain size associated with larger casting sections lead to some decrease in tensile strength, fracture toughness and fatigue behavior.

Table: Average prior austenitic grain size for quenched and tempered grades as a function of the section size.

	5"	3"	1"
LCC	4.6	5.6	6.3
LC1	4	6	7
LC2	4	4.5	5.5
LC3	4	5	5.8
CA15	3	4	4
CA6NM	3	3	3.6
Total Average	3.7	4.7	5.4

The microstructure of a steel casting can normally be refined by heat treatment. Heat treatment can produce finer microstructures than the as cast microstructure. However, this finer microstructure depends on the cooling rate from the austenitizing temperature. In thicker sections it is impossible to cool the center of a casting as quickly as the edge. The finer microstructure nearer the surface gives better mechanical properties as shown in figures.

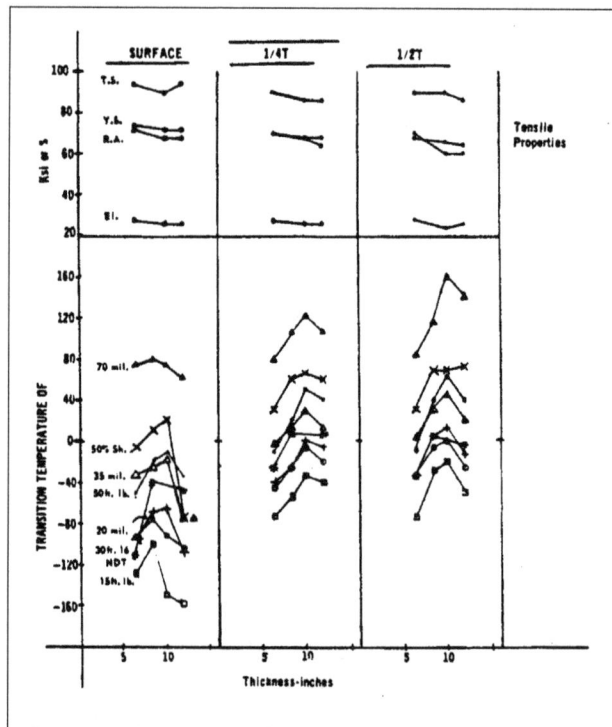

Surface to center variation of tensile and impact properties in 5 to 10 inch sections.

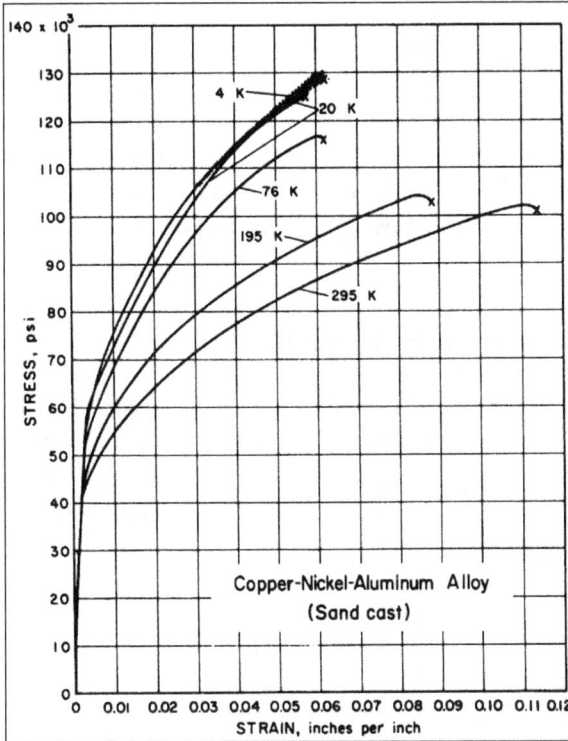

Distribution of endurance limit across the thickness of steel casting of several section thickness.

In figure above, sections of about 5 to 10 inches were tested for tensile and impact properties. The variation from surface to center is shown and; as expected, the tensile strength is less in the center and the transition temperature is higher.

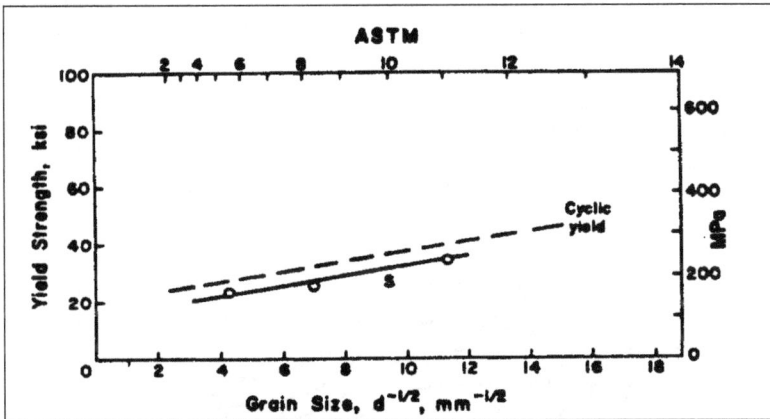

Effect of grain size on endurance limit of a low-carbon steel.

In the 1030 specimens, normalized and tempered, the properties are fairly insensitive to section size up to 6 inches, with the endurance limit being about 37,000 psi. With the 8630 material, normalized and tempered, the endurance limit improved with a value of 44,000 psi in the center of the 1-1/4" thickness. The 8635 material, quenched and tempered, had endurance limits of 54,000 psi, 48,000 psi, and 38,000 psi in the center of the 1-1/4", 3", and 6" thickness. The section size variations are the most pronounced in the quenched and tempered condition; since quenching cannot always extract the heat in the center of a thick section fast enough to form martensite.

The single most important and least avoidable effect of section size is the coarseness of the microstructure, since the cooling rate at the center of thick sections will never be rapid. Intercritical heat treatment might allow some refinement of the microstructure in thick sections.

Casting Discontinuities and Defects

Casting larger section sizes can agravate a number of casting discontinuities like the larger inclusion sizes reported in table. The larger inclusion sizes do not have much of an effect on tensile strength but do lower the impact strength, in figure below, throughout the range. Larger inclusions also lower the fatigue resistance, in figure below, particularly in higher strength steels.

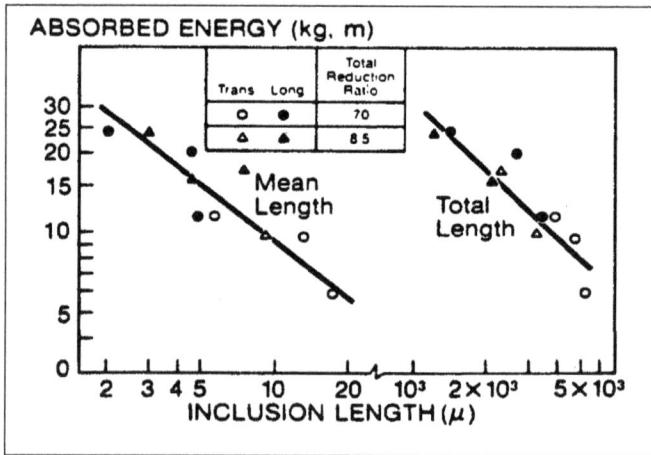

Effect of inclusion length on absorbed energy in the longitudinal and transverse direction.

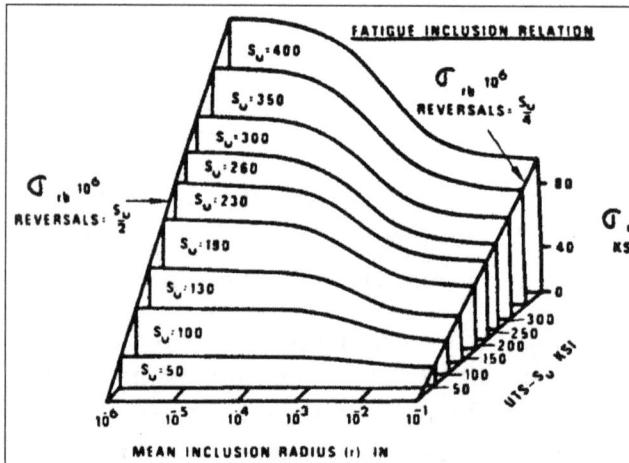

Effect of inclusion size on fatigue $\left(\sigma_{rb}\right)$ as a function of ultimate tensile strength (S_u).

Table: Effect of the section size on the average length of Type II inclusions.

| (Length of Inclusion in Microns) | | | | |
	1"	3"	5"	Average % Sulfur
WCA	80	80	80	.022
WCB	70	150	295	.018
WCC	180	350	700	.035
LCC	445	460	550	.030
LC2	66	230	265	.018
LC3	105	145	270	.021
WC6	70	180	235	.015
WC9	40	80	525	.025
Total Average	132	209	365	

One source of embrittlement agravated by larger section sizes is aluminum nitride, or "rock candy" fractures. Above figure illustrates the decreasing tolerance for aluminum and nitrogen with a decreasing cooling rate. This concern and inclusion type control shows the need for a well thought out and tested deoxidization practice for heavy section castings.

Another problem agravated by thick sections is temper embrittlement. Temper embrittlement is caused by the segregation of impurities such as phosphorus, arsenic, antimony, and tin into the grain boundary areas. The embrittlement shows as an upward shift of the transition temperature after exposure to temperatures in the range of 750- 1100° F. Temper embrittlement can occur in large sections during cooling from the tempering. Cooling from welding can also induce temper embrittlement.

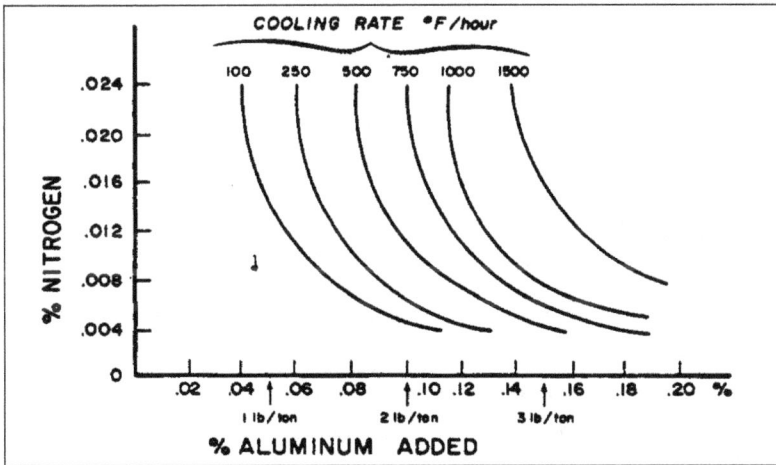

A chart to indicate approximate limits of nitrogen and aluminium that may be tolerated in a base analysis containing 0.30% C, 1.60% Mn, 0.50% SI, 0.50% Cr, and 0.35 % Mo without development of intergranular type fracture.

Other defects are more prone to happen in thick sections such as microshrinkage, surface roughness, and surface contamination. These defects can also cause some deleterious effects on the properties of steel castings.

Geometrical Size Effect

The geometrical size effect is measured as the difference in properties obtained when different sized specimens of similar metallurgical background are tested. This size effect has been investigated for tensile, impact and fatigue properties. In general, the mechanical properties of a steel are not as favorable in the larger size specimens. This decrease in properties with larger specimens had been explained by greater probability of favorable grain orientation on the surface or larger flaws when a greater amount of material is tested. While this statistical explanation does offer some rationale for the poorer properties found in larger specimens, fracture mechanics offers a more satisfying explanation.

Fracture Mechanics

Linear Elastic Fracture Mechanics (LEFM) was developed to explain the failures of brittle materials in the presence of defects at stresses well below the strength of the material. LEFM was subsequently extended to explain the behavior of steels especially high strength steels in the presence of a flaw. The beauty of LEFM is the use of one variable that relates load, flaw size, and part configuration to failure. This allows the use of LEFM test results to be used in design to prevent brittle type failures. The most widely used variable use to characterize LEFM behavior of materials is K_{IC}. Other variables have been used such as G and J_{IC}. J_{IC} has some advantages over K_{IC} in lower strength steels since it was developed for material exhibiting larger amounts of yielding in failure; however, K_{IC} is the most common measurement.

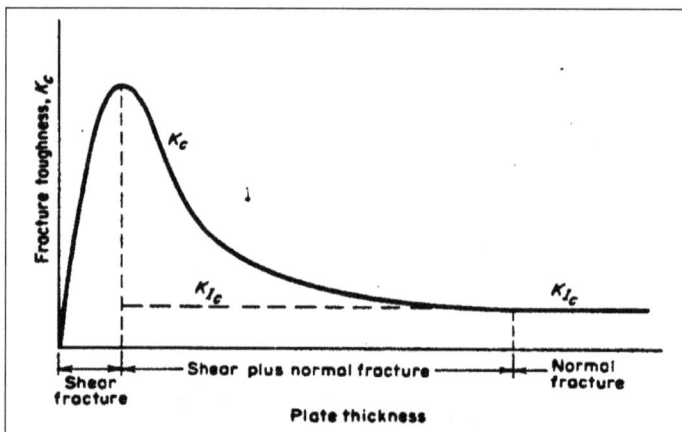

Schematic variation of the fracture toughness as a function of the plate thickness.

K_{IC} relates stress and flaw size and has the units, ksi/in. The relationship of K_c, the critical stress intensity for static loading, to the plate thickness is shown in figure. As the plate thickness increases, the plastic constraint increases until a plane strain condition exists; and the K_c value decreases to the K_{IC} value. Once the K_{IC} value has been determined, it can be used as a material property in design. Because of the effect of increasing section size, increasing the plastic constraint and establishing plane strain conditions, it should not be surprising that thicker sections fail in the more brittle manner explained by LEFM, rather than by tensile - yield relationships.

Fatigue

When a stress is applied to a steel part and the existing defects are below critical size, the part will not fail on a single application of the load. However, if the load is repeated, a crack can initiate from existing defects, grow, and finally induce failure. The final failure occurs in a manner previously described in LEFM, but the process of crack initiation and growth under repeated loads is known as fatigue. Fatigue can be discussed as crack initiation and crack growth. Crack growth can be examined as growth rate (da/dN) for repeated applications of a load (K range of stress intensity). In figure, there are fatigue crack propogation rates for six steels. The fatigue crack initiation or growth rates are not very affected by section size. However, heavy sections with their increased plastic constraint are less resistant to the final failure from the fatigue crack.

Cast Iron Mechanical Properties

Cast iron has many good mechanical properties, these are given below:

Tensile Strength

Tensile strength is the most important physical property for iron castings. Its unit is N/mm². Tensile strength will show the strength of cast iron when it is pulled until broken. For example, the ductile iron 60-40-18 has tensile strength of 60000 psi, therefore, it is equal to 414 Mpa. Therefore, if your parts need to use this material, then your supplier should keep their iron castings with min. 414 Mpa tensile strength.

Yield Strength

Yield strength is another definition to define the strength of cast iron. It is not very necessary to check this property since it just more deeply measure the strength of cast iron. The tensile strength has already covered the most of requirements to the strength. As for the ductile iron 60-40-18, its yield strength is 40000 psi, equal to about 276 Mpa.

Elongation

Sometimes, elongation is very important. High elongation means the iron casting can be pulled to longer, which means they have better ductility or called as malleability.

Therefore, they will not be broken easily. The ductile iron has higher elongation than gray iron. As for the ductile iron 60-40-18, it has 18% elongation.

Hardness

Hardness is also important sometimes. As for the machined parts, the hardness should not be too high to affect the machining process. As usual, the Brinell Hardness from 160-220 should be a good range. Too high may cause the hard spots during machining, too low will affect the application.

Tensile Properties of Aluminum Foundry Alloys

The tensile properties of an alloy can be exploited if detrimental defects and imperfections of the casting are minimized and the microstructural characteristics are optimized through several strategies that involve die design, process management and metal treatments.

The foundry industry has constantly tried to address the challenge of producing high quality and cost effective castings as the final applications demand stringent conditions. There have been constant efforts to minimize defects and imperfections in the castings and to optimize the microstructure, keeping in mind the main variables related to the employed alloy, the initial melt quality and the process conditions.

From the design viewpoint, the knowledge of tensile properties of cast products is a relevant topic, which is not fully covered by existing international standards, except for the newly introduced document CEN/TR 16748:2014. As a matter of fact, the European standards on foundry Al alloys (EN 1676 and EN 1706) provide interesting information, which should be integrated with more recent studies. Particularly, the EN 1706 standard specifies the chemical composition limits for Al casting alloys and their tensile properties. As specified in, the yield strength (YS), the ultimate tensile strength (UTS) and the elongation to fracture (EL%) in as-die-cast condition are, respectively, 140 MPa, 240 MPa and less than 1% for the most high-pressure die-cast (HPDC) alloys, while these are around 90 MPa, 170 MPa and in the range 1–2.5% for the main gravity die-cast (GDC) alloys. Hence, the EN 1706 standard reports minimal and conservative values of the tensile properties, by considering that the average content of defects and imperfections in Al alloy castings could worsen their static strength. Recently, a consortium called StaCast, comprised of various institutions from European countries, was constituted and successfully proposed in, which should be used together with the existing standards for the evaluation of Al alloy tensile properties. In general, the CEN/TR shows higher values of tensile properties for Al alloys based on studies that optimize and improve die parameters, by consequently minimizing defects and imperfections and by improving microstructure in the castings.

The variables affecting the tensile strength of cast Al alloys are certainly well articulated and complex, involving casting and testing issues. Indeed, the static strength of foundry alloy components mainly depends on alloy composition, casting conditions, heat treatment, geometry of separately poured specimens and test conditions. Since the effects of alloying and heat treatment do not belong to the scope of the work, this presents analysis and comparison of the tensile properties measured on the separately poured specimens obtained using dies and process parameters that were recently made part of in order to explore the tensile strength of die cast Al-foundry alloys. The most popular Al foundry alloys were identified through a questionnaire: about 60 foundries, which cover a consistent percentage of the European production, have answered to the questionnaire. It was found that the most common alloy category is comprised of Al-Si based alloys, both for HPDC and GDC processes. For instance, StaCast consortium results revealed that the $AlSi_9Cu_3(Fe)$ alloy is used by around 59% of European foundries, while both $AlSi_7Mg0.3$ and $AlSi_{12}Cu_1(Fe)$ alloys are used by around 35%. Moreover, 31% of foundries employ $AlSi_{11}Cu_2(Fe)$ and a percentage lower than 30% employs other alloys, such as $AlSi_{12}(Fe)$.

Reference Castings for Testing Tensile Behavior

A reference die is a permanent mold, designed according to the state-of-the-art methodologies and made of steel or cast iron, suitable for the evaluation of the static strength of a cast alloy. The geometry of such a die varies in accordance with the applied kind of process, i.e., HPDC or GDC. The specimens manufactured using such a reference die is called separately poured specimens.

As for process, the casting parameters affecting the quality of components were deeply reviewed in previous works for HPDC and GDC. It was observed that, by modifying the existing standard permanent molds, higher tensile properties could be obtained from conventional Al foundry alloys. Numerical simulation studies have been quite effective in understanding the distribution of porosity during melt flow and optimizing die parameters. Regarding mechanical test conditions, capacity of machine, sensor system, cross-head speed, data elaboration, time between manufacturing and test, and testing temperature affect the behavior of castings. Tensile specimens can be round or flat, machined or not, and are characterized by a specific geometry (i.e., gauge length, gauge width and radius).

Reference Castings for High-pressure Die-cast Al-Si Alloys

The static strength of HPDC Al-Si alloys can be evaluated by the reference die designed, built and tested in the frame of the NADIA Project. The reference casting #1, shown in figure, is suitable for various kinds of characterization, as it has round fatigue and stress-corrosion bars, corrosion-Erichsen test plate, fluidity and Charpy test appendices, besides the flat tensile bars. The dimensions of round and flat tensile specimens are shown in figure.

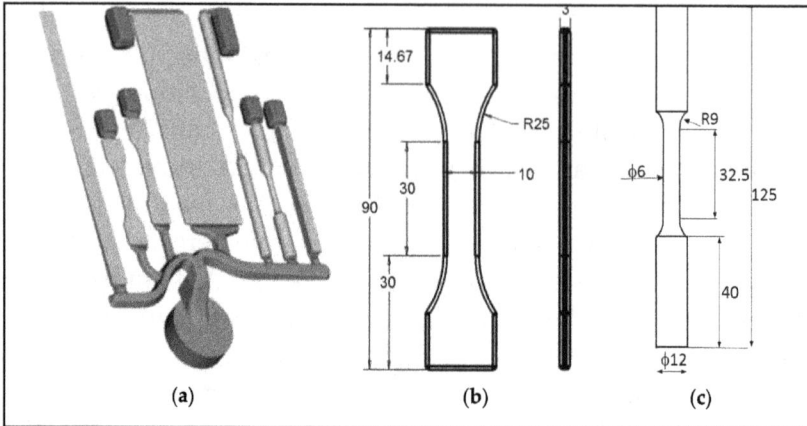

(a) Reference casting #1 for high pressure die casting; (b) flat; and
(c) round tensile specimens (dimensions in mm).

This reference die is made by two AISI H11 parts, a fixed side and ejector side, and is shown in figure. The die was carefully designed to obtain a uniform molten metal front and a favorable thermal evolution inside the die cavity. Figure shows some results of the numerical simulation of the filling process, which provides evidence of the melt velocity distribution inside the die cavity at three different percentages of the die cavity filling.

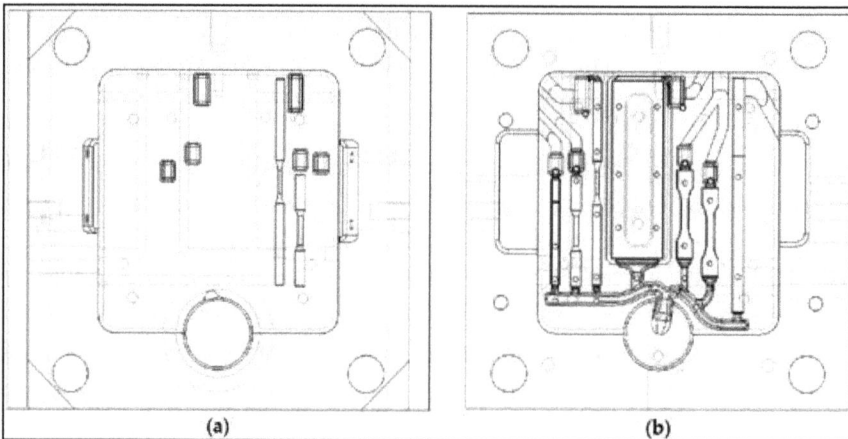

Layout of the die for diecasting of specimens: (a) fixed side and (b) ejector side.

Timelli et al. successfully used this reference die to determine the tensile properties and the microstructural characteristics of diecast $AlSi_9Cu_3(Fe)$ alloy. It was observed that the tensile specimens from the reference casting #1 show significantly higher tensile strength and elongation as compared to machined specimens from the same casting. This is related to heterogeneity in the microstructure from the periphery to the center of the cross section of the die-cast specimens. It is reported that there exists a surface layer of about 1.3-mm thick in the tensile bars that is free of porosity, which contributes to higher mechanical properties in die-cast specimens if compared to machined ones.

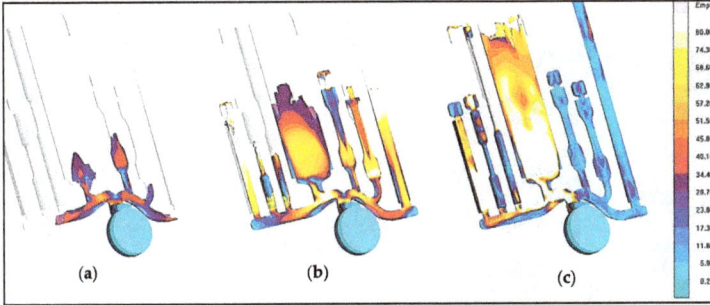

Calculated melt velocity at (a) 44%; (b) 66%; and (c) 92% of
die filling. The color code indicates velocity in m s⁻¹.

The reference casting #1 has been also adopted for the experimental validation of an analytical method for explaining and predicting the quality of castings, by means of the root means square plunger acceleration and the plunger speed extracted from the plunger displacement curve.

The mechanical behavior of Al foundry alloys can also be estimated by means of the reference die designed, built and tested by HYDRO in cooperation with NTNU (University of Science and Technology, Trondheim, Norway). The dimensions of the cylindrical tensile test specimens are shown in figure. Some previous works demonstrated that the tensile specimens show higher porosity in the grip section as compared to the gauge section. Moreover, the findings indicate that the central region of the grip section solidifies later than both the surface region and the gauge section. Similar observations to those made for the tensile specimens from the reference casting #1 have been proposed by Timelli et al. for the reference casting #2.

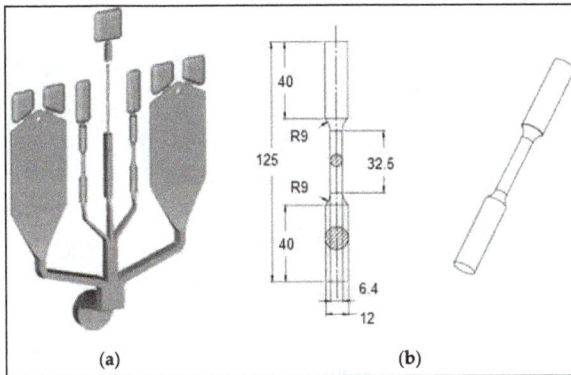

(a) HPDC reference casting #2; (b) round tensile bar (dimensions in mm).

Reference Castings for Gravity Die-cast Al-Si Alloys

Gravity casting, being the simplest and conventional route of casting, has few disadvantages such as entrained air, bubbles and oxide bi-films. The design of sprue, runner and gating system is critical in minimizing defects and imperfections. It can be achieved by critical calculations, like those proposed by Campbell.

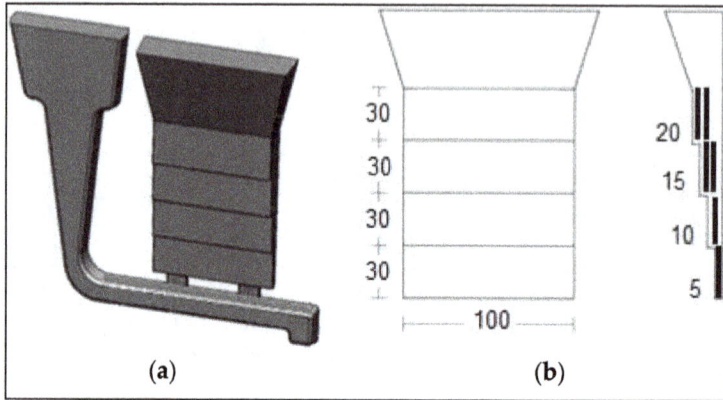

(a) Reference casting #3 for gravity die casting; (b) sectioning
scheme for mechanical property testing (dimensions in mm).

The ASTM B108 gives dimensions and details of a standard round tensile specimen specifically designed for GDC, which is also called a Stahl mold. Efforts made by researchers and foundries to adopt the Stahl mold for obtaining the best mechanical properties of cast alloys are reviewed in . The limitation of the Stahl mold was that it has well-defined dimensions in terms of gauge length and thickness, with the consequence that it was not allowed to study different solidification rates. To overcome this limitation, the Aluminum Association developed a step configuration of the die with variable thickness. The design proposed by Grosselle et al., shown in figures as reference casting #3, consists of four steps varying from 5 to 20 mm, from which flat tensile bars can be machined as per the ASTM B557 with gauge length, width and thickness of 30, 10 and 3 mm, respectively.

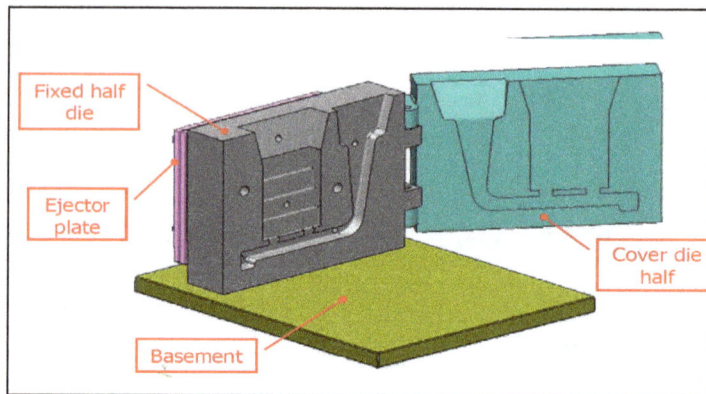

Layout of the die.

The step casting was gated from the bottom of the thinnest step, while the riser over the casting ensures a good feeding. This configuration allows for obtaining a range of cooling rates and consequently different microstructural scales in the casting. As shown in figure below, the two-part die is split along a vertical joint line passing through the pouring basin. To facilitate assembly and mutual location, the die halves are hinged. The dimension of the whole die is $310 \times 250 \times 115$ mm^3 and the thicknesses of the two die halves are 45 and 75 mm, respectively.

Variations of the step design have been recently studied to improve the quality of test specimens obtained. It is worth noticing the optimization of the step mold design proposed by Timelli et al. for Mg alloys, which could be probably used for other foundry alloys.

The mechanical behavior of GDC Al-Si alloys can be evaluated by another step reference die in cooperation with SINTEF (Trondheim, Norway). As shown in figure, the reference casting #4 has five steps, 250 mm length, 120 mm width and thickness ranging from 5 to 30 mm. Round tensile bars can be machined from steps with 30, 20, 15 and 10 mm thickness, while flat bars from the 5 mm step. The round bars have 36 mm gauge length and 10 mm total diameter, while the flat bars have 32 mm gauge length, 10 mm total width and 5 mm thickness.

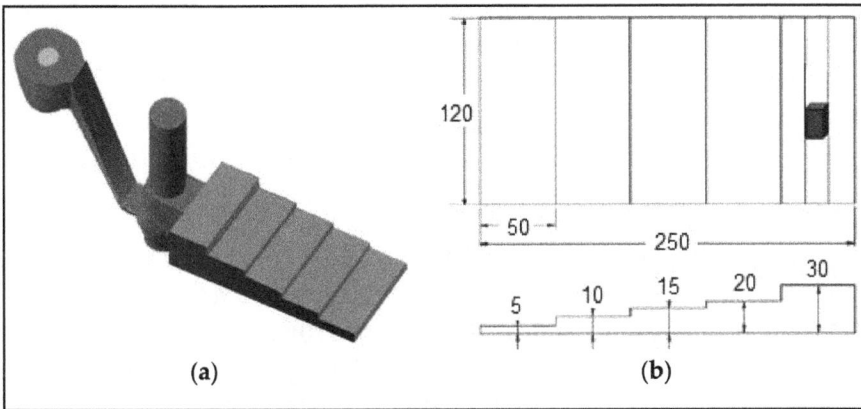

(a) Reference casting #4 for gravity die casting; (b) sectioning scheme for mechanical property testing (dimensions in mm).

Results on the Expected Tensile Strength of Al-Si Alloys Cast in Permanent Mold

The expected tensile strength is the mechanical behavior that can be achieved by Al-Si alloys, cast in reference dies with state-of-the-art knowledge on die design, process management and alloy treatments properly applied to minimize casting defects and imperfections and to improve the microstructure. The expected tensile strength of Al-Si alloys was estimated by means of tensile testing performed on specimens obtained through the above-described dies.

Expected Tensile Strength of High-pressure Die-cast Alloys

The tensile strength of HPDC alloys was been obtained through reference casting #1. Table collects the chemical composition of the alloys tested, which have been selected based on the considerations made in the Introduction. The addition of iron permits avoiding soldering phenomenon due to high velocity and pressure typical of the HPDC process.

Table: Chemical composition of the investigated high pressure die cast Al-Si alloys (wt. %).

Alloy	Si	Fe	Cu	Mn	Mg	Cr	Ni	Zn	Pb	Sn	Ti	Al
Al-Si$_9$Cu$_3$(Fe)	8.227	0.799	2.825	0.261	0.083	0.083	0.081	0.895	0.083	0.026	0.041	bal.
Al-Si$_{11}$Cu$_2$(Fe)	10.895	0.889	1.746	0.219	0.082	0.082	0.084	1.274	0.089	0.029	0.047	bal.
AlSi$_{12}$Cu$_1$(Fe	10.510	0.721	0.941	0.232	0.045	0.045	0.080	0.354	0.025	0.038	0.038	bal.

The alloys, supplied as commercial ingots, were melted in a 300 kg crucible in a gas-fired furnace set up at (800 ± 10) °C and maintained at this temperature for at least 3 h. The temperature of the melt was then gradually decreased by following the furnace inertia up to (690 ± 5) °C. The molten metal was degassed with Ar for 15 min. Since the initial metal quality can deeply affect the castability of an alloy and the final properties of castings, the quality of the materials used in the experimental campaigns was estimated by Foseco H-Alspek to measure hydrogen level and by reduced pressure test (RPT) to measure bi-film index. The hydrogen content was lower than 0.15 mL/100 g Al during the entire experimental campaigns while the bi-film indexes were in the range between 10 and 18 mm.

Cast-to-shape specimens were produced using HPDC reference casting #1 in a cold chamber die-casting machine with a locking force of 2.9 MN. The nominal plunger velocity was 0.2 m/s for the first phase and 2.7 m/s for the filling phase; a pressure of 40 MPa was applied once the die cavity was full. The optimal experimental conditions at steady-state, which guarantee a high quality of castings, were found to be: pouring temperature 690 °C, melt velocity at in-gates 51 m/s and filling time 9.7 ms. Indeed, the process parameters influence defect content and microstructure of castings, as it has been deeply studied in. The cycle time was approximately 45 s.

The weight of the Al alloy die-casting was 0.9 kg, including the runners, gating and overflow system. About 15 castings were scrapped after the start-up, in order to reach a quasi-steady-state temperature in the shot chamber and the die. Oil circulation channels in the die served to stabilize the temperature (at ~230 °C). The melt was transferred in 18 s from the holding furnace and poured into the shot sleeve by means of a coated ladle. The fill fraction of the shot chamber, with 70 mm inner diameter, was 0.28.

The surface finish of samples was adequately accurate to avoid machining, and only some excess flash along the parting line of the die was manually removed. The tensile tests have been done on a tensile testing machine. The crosshead speed used was 2 mm/min and the strain was measured using a 25-mm extensometer. Experimental data have been collected and processed to provide yield stress (YS or 0.2% proof stress), ultimate tensile strength (UTS) and elongation to fracture (%EL). At least 10 tensile tests were conducted for each condition. The specimens were maintained at room temperature for five months before testing. Table summarizes the tensile strength of the alloys tested.

Table: Tensile strength properties of the investigated Al-Si HPDC alloys obtained through reference die #1. Flat specimens with 3 mm thickness and round specimens with 6 mm diameter.

Alloy	Type of Specimen	UTS (MPa)	YS (MPa)	EL (%)
$AlSi_9Cu_3(Fe)$	Flat Round	309 ± 6	163 ± 1	3.6 ± 0.3
		342 ± 8	168 ± 6	5.1 ± 0.4
$AlSi_{11}Cu_2(Fe)$	Flat Round	312 ± 2	153 ± 1	3.5 ± 0.1
		342 ± 7	153 ± 3	5.5 ± 0.7
$AlSi_{12}Cu_1(Fe)$	Flat Round	383 ± 2	137 ± 1	3.5 ± 0.1
		315 ± 7	131 ± 2	7.1 ± 0.5

Tensile properties of the round specimens obtained through $AlSi_9Cu_3(Fe)$ alloy can be compared with those achieved in a previous study using the same die. The values of mechanical properties are in good agreement, by highlighting effectiveness of the proposed die in evaluating the properties of castings.

Properties reported in table can be also compared with those obtained in another study, which used a different geometry called "one bar casting" with a single large feeding channel on one extremity of the specimen, thus to have a coaxial inflow with the specimen. Moreover, a large feeder was added on the other extremity of the specimen. The resultant as-cast tensile bar was cylindrical with gauge length 30 mm, total length 96 mm and gauge diameter 8 mm. For the $AlSi_{10}Cu_3(Fe)$ alloy, the following properties were achieved: 275 MPa UTS, 150 MPa YS and 2.1% EL. Besides its effectiveness, another advantage of the proposed die is that multiple specimens for tensile, fatigue, impact, corrosion and stress-corrosion testing can be prepared from a single casting.

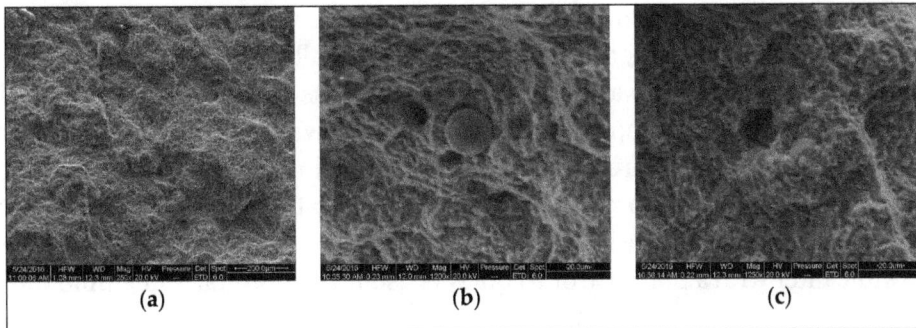

SEM micrographs of the fracture surface of the round specimen obtained through reference die #1 and $AlSi_{12}Cu1(Fe)$ alloy: (a) low magnification image; (b) cold shot and (c) micro-porosity. These defects are small thanks to the optimized geometry of the die.

This reference die was carefully designed and optimized to maximize the quality of castings, by reducing the scrap percentage and the presence of defects and imperfections. With the aim of demonstrating this statement, a typical fracture surface under the scanning electron microscope (SEM) is reported in figure, and some sample defects detected in the specimen at higher magnification are also shown. These defects are detrimental for tensile properties and could cause premature failure of castings. Nevertheless, they are small thanks to the optimized geometry of reference die and the optimal experimental conditions adopted. More on casting defects and their effect on mechanical properties could be found in compendium.

Expected Tensile Strength of Gravity Die-cast Alloys

The tensile strength of GDC alloys was evaluated through reference castings #3 and #4. Table collects the composition of the alloys tested. These alloys were selected based on the popular Al foundry alloys identified through a questionnaire and are frequently used for manufacturing automotive components, such as cylinder heads, wheels and carter. The chemical composition of the alloys was properly chosen for improving the final properties of castings. For instance, the addition of Cu permits enhancing strength and workability of an alloy, while the presence of Fe improves wear resistance.

Table: Chemical composition of the investigated gravity die cast Al-Si alloys (wt. %).

Alloy	Si	Fe	Cu	Mn	Mg	Ni	Zn	Ti	Al
$AlSi_7Mg0.3$	6.5	0.1	0.002	0.007	0.3	0.003	0.006	0.1	bal.
$AlSi_6Cu_4$	6.0	1.0	4.0	0.5	0.1	0.3	1.0	0.2	bal.

The alloys, supplied as commercial ingots, were melted in 70 kg electric resistance furnace set up at $(720 \pm 10)\,°C$ and maintained at this temperature. The molten metal was then degassed with Ar for 15 min. The hydrogen content of the melt in the holding furnace was analyzed by hydrogen analyzer, and it showed values lower than 0.1 mL/100 g Al during the entire experimental campaign. Periodically, the molten metal was manually skimmed with a coated paddle. Sr modification was carried out as well as Ti grain refining; lower mechanical properties can be expected without these metal treatments. Castings were produced using reference castings #3 and #4. The temperature of the die was maintained at around 300 °C by means of oil circulation channels, and about three castings were scrapped after the start-up to reach a quasi-steady-state temperature in the die. A ceramic filter with a pore size of 10 ppi was used. The optimal filling time was 5–6 s.

Flat tensile test bars with rectangular cross section were drawn from each step, in the middle zones of the castings and the dimensions were maintained as per the ASTM-B557 standard. The tensile tests have been done on a tensile testing machine with crosshead speed of 1.5 mm/min. The strain was measured using a 25-mm extensometer. Table below summarizes the tensile strength of the investigated alloys. Several specimens (around 10) from each step were tested.

Table: Average tensile strength of Al-Si GDC alloys evaluated by means of two different reference die.

Alloy	Step Thickness (mm)	Reference Die #3		Reference Die #4	
		UTS (MPa)	EL (%)	UTS (MPa)	EL (%)
AlSi$_7$Mg0.3	5	181	2.3	194 ± 2	9.5 ± 1
	10	172	1.7	182 ± 3	7.1 ± 0.5
	15	182	2.3	174 ± 1	5.6 ± 0.3
	20	167	1.8	166 ± 2	4.4 ± 0.2
	30	-	-	161 ± 3	3.2 ± 0.6
AlSi$_6$Cu$_4$	5	203	0.9	-	-
	10	207	1.0	-	-
	15	194	0.8	-	-
	20	188	0.7	-	-

The AlSi$_7$Mg0.3 alloy is equivalent to the popular A356 cast alloy and the minimum standard tensile values of this alloy reported in the literature are 145 MPa (UTS) and 3% EL. The AlSi$_6$Cu$_4$ is equivalent to A319 alloy whose minimum strength values reported in the literature are 186 MPa (UTS) and 2.5% EL, measured using the popular ASTM B-108 mold. Considering the values reported in the values of Table for reference die #3 are disappointing and offer scope for the improvement in the mold design. Generally, low elongation to fracture is related to the presence of diffused microscopic casting defects such as oxides, porosity, etc. For similar alloy composition, Akhtar et al. observed higher ductility using reference die #4. The comparison between reference dies #3 and #4 indicates that the die configuration is an important parameter in assessing the tensile potential of an alloy as much as the melt quality and the pouring conditions. Timelli et al. noted that by modifying the runner and gating systems in reference die #3, the amount of casting defects could be minimized. This was shown in both experimental and numerical simulation studies. The modified die design is represented in figure. However, the study was limited to microstructural characterization of magnesium alloys and needs to be extended to aluminum alloys. By means of numerical simulation techniques, Wang et al. found that the gauge section of the standard Stahl mold (standard ASTM B-108) showed higher porosity (about 3%) compared to the AA Step mold (less than 1%) in an AlSi$_7$Mg alloy. This comparison confirmed the statement of Singworth and Kuhn that the Stahl mold could not produce better mechanical properties than the Step mold, due to higher micro-porosity in the gauge section on account of micro-shrinkage. However, the AA Step mold is different from reference die #3 in terms of in-gate and runner design. To the best of the authors' knowledge, there is no research available about the comparative studies between the Stahl mold and the modified Step mold.

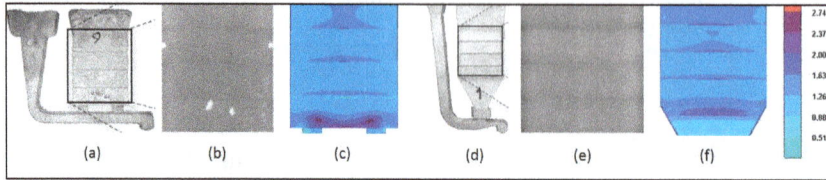

(a) step casting produced with reference die #3; (b) defects observed under X-ray scan; (c) distribution of micro-porosity due to shrinkage predicted by magma-simulation; (d) modified step casting; (e) improved quality of casting as observed under X-ray scan; (f) distribution of micro-porosity percentage (%) as indicated by the color code.

Figure shows the microstructure of the $AlSi_7Mg0.3$ alloy as a function of step thickness. It is worth noticing that the geometry of the proposed reference die permits accurately evaluating the mechanical properties of the alloy with thickness variations. As the step thickness reduces, the cooling rates are higher and secondary dendritic arm spacing (SDAS) is lower, resulting in improved mechanical properties. Average SDAS of each step has been measured and was 45.4 µm, 32.4 µm, 30.1 µm and 25.0 µm by reducing the thickness. Percentage of porosity has also been estimated and was 0.65, 0.42, 0.22 and 0.05 by reducing the step thickness. Similar observations were made in the work of Grosselle et al. with reference die #3, which is concomitant with observations made using reference die #4 for similar alloy composition.

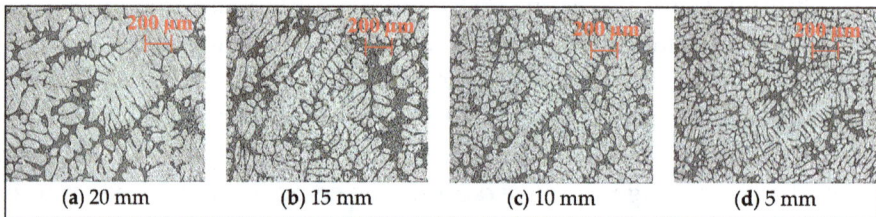

Microstructure of the $AlSi_7Mg0.3$ alloy with decreasing step thickness from (a–d). These micrographs are related to interior of the casting.

Fracture Toughness of Metal Castings

From the continuum mechanics point of view, fracture toughness of a material may be defined as the critical value of the stress intensity factor, the latter depending on a combination of the stress at the crack tip and the crack size resulting in a critical value.

The local stress σ_{local}, shown in figure, scales as $\sigma\sqrt{(\pi c)}$ for a given value of r, where σ is the remotely applied stress, σ_{local} is the stress in the vicinity of the crack at a distance r from the tip and c is the crack length; this combination is called the Stress Intensity Factor (K). For the type of load shown (tensile load) K is denoted as K_I. Thus,

$$K_I = Y\sigma\sqrt{(\pi c)}$$

where Y is a dimensionless constant to account for the crack geometry. K_I has units of MPa $m^{1/2}$. The material fractures in a brittle manner when K_I reaches a critical value, denoted by K_{IC}; if there is significant crack tip plasticity, instability occurs at this critical value, leading to fracture. A simpler view of the fracture toughness is that it is a measure of the resistance of the material to separate under load when a near-atomistically sharp crack is present.

The stress intensity factors are usually identified by the subscript I for "opening mode", II for "shear mode" and III for "tearing mode". The opening mode is the one that has been investigated widely and hereafter only K_I will be considered.

Under load, a metallic material first undergoes elastic deformation and plastically deforms when its yield stress is exceeded. Fracture occurs when the ability to plastically deform under load is exhausted. The chief cause of the plastic deformation is the movement of dislocations and the resistance to its movement causes increased plastic flow stress and abetment of fracture.

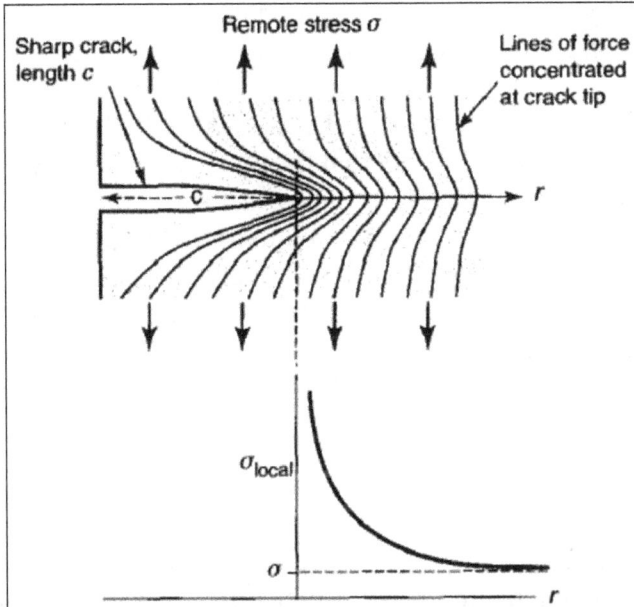

Lines of force and local stress variation from a body with sharp crack.

The term "metal casting" represents an umbrella consisting of many variants. The composition of the metal (alloy) will usually depend on the variant. The common factor among the variants is that they are all products of liquid-to-solid transformation, usually termed "solidification". Solidification of castings involves nucleation and growth of solid. Casting alloys usually consist of more than one phase. The simplest solidification occurs in a pure metal or an isomorphous system wherein the solid consists of only one phase. As the complexity increases, an eutectic system consisting of two solid phases may be formed, which may be totally different in properties. In low carbon steel castings, a high temperature reaction known as peri-tectic reaction will occur, which have some influence on the room temperature mi-crostructure. Adding to this complexity, solid-to-solid transformations may occur, as for example, the eutectoid reaction in cast iron and steel. The phase diagrams will at best give useful guidance on the development of microstructure as they are based on equilibrium, but most castings solidify under nonequilibrium conditions resulting in departures from the phase diagram predictions. Commercial castings invariably contain various impurities that may affect the microstructure. Certain aspects of the casting microstructure have a fundamental influence on the fracture resistance.

It is therefore pertinent to consider the influence of valence electrons on the frac-ture behavior. Covalent bonds have shared electrons and the limited mobility of the electrons impedes plastic flow resulting in brittle fracture. Though metal castings in general have metallic bonds, they may contain covalent compounds such as nitrides, carbides and others as inclusions. Silicon, an important constituent in Al-Si casting alloys also has covalent bond. Ionic bonds permit better electron mobility than co-valent bonds, but may cause brittle behavior when like poles interact while slipping. Metallic bonds offer least restriction to electron mobility, but as stated earlier, only a few commercial castings are made of pure metals or isomorphous alloys. Another important factor that needs attention is the dislocation dynamics as affected by the casting microstructure, despite the fact that the dislocation density in castings is much lower than in cold worked materials, in the as-cast state; this difference may get less under stress in castings. Unfortunately little quantitative information is available on the significance of dislocation dynamics on fracture in castings. Some casting alloys have compositions suitable for heat treatment involving solid state transformations. The microstructure is substantially changed after heat treatment and thus, in heat treated castings, the fracture behavior will usually be different as compared to the as-cast counterparts.

From the microstructural point of view, the route to increase the fracture toughness of castings would involve conflict in increasing both the fracture toughness and the yield strength. The factors of importance are: improved alloy chemistry and melting practice to remove or make innocuous impurity elements that degrade fracture toughness; de-velopment of microstructures and phase distributions to maximize fracture toughness. Through proper choice of composition and process variables; microstructural refine-ment through solidification control.

Thus it is clear that continuum mechanics provides the theoretical basis for designing against fracture in castings, but a thorough understanding of the microstructure and its effects is essential to fine-tune the final design against fracture. As noted by Ashby, the real value of a well-functioning product is easy to assess, but the value of a failed product eludes evaluation until the extent of damage is known. Such knowledge can often fall under the category of "too little, too late".

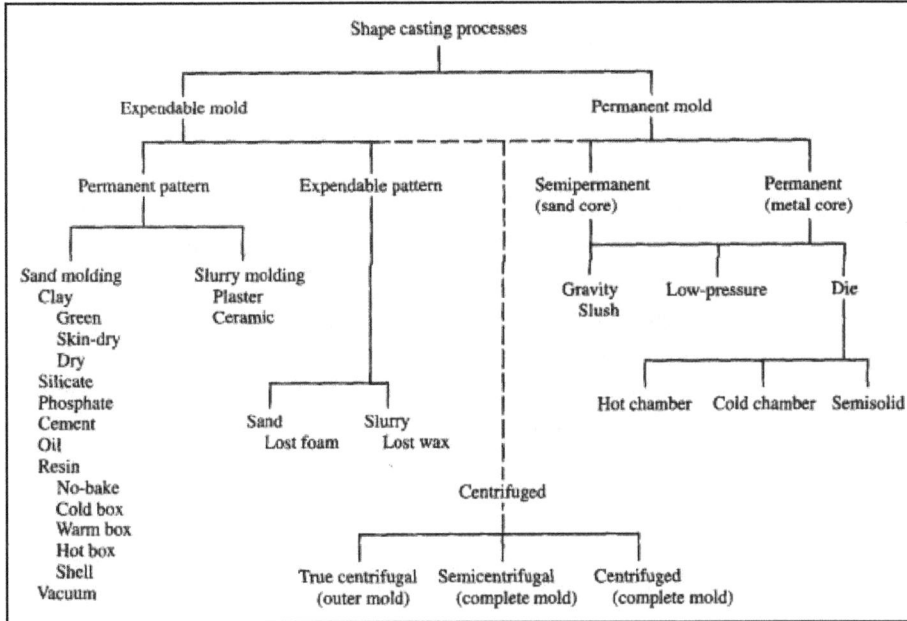

Classification of Metal Casting Processes.

Casting Processes

The umbrella covering the casting processes is shown in the figure. It is clear from figure that there are many avenues for making a casting, depending on the type of pattern, the type of mold and whether pressure is used for assistance in filling the mold. Not all processes are suitable for all the casting alloys. Investment casting (ceramic slurry, lost wax) is perhaps the most accommodative process for most alloys and others have limitations based on resistance to high temperature, chemical reaction and other factors. It is therefore customary to choose the casting process with due regard to the casting alloy. A recent addition to the umbrella is the squeeze casting process which is somewhat analogous to transfer molding of polymers. The microstructure of the casting is strongly affected by the process used for making it.

As stated earlier, casting is the product of solidification, which consists of nucleation and growth of solid from the liquid metal alloy). The final microstructure is decided by the composition of the alloy, the solidification rate and any melt treatment used. The alloys, based on their phase diagram may be of long-freezing range or short-freezing range type.

The solidification rate is governed by the rate at which the mold is able dissipate the latent heat and superheat of the metal poured into the mold. Permanent molds like metal and graphite molds have higher thermal conductivity than disposable molds like sand and ceramic shells and therefore provide higher solidification rates. If there is no melt treatment, finer scale microstructure can be expected when these higher conductivity molds are used. Melt treatment however, can change this picture. The object of this treatment is to refine the microstructure and the treatment is variously termed as grain refinement (in the case of single phase alloys), modification or inoculation (in the case of second phase alteration of binary alloys).

The application of continuous pressure as in squeeze casting may also substantially affect the microstructure. Long-freezing range alloys cooled at a relatively slow rate, as for instance in a sand mold, tend to solidify in a "mushy" or "pasty" manner. During the progress of solidification, there will be three distinct zones: liquid, liquid+solid, solid in most cases. The liquid+solid zone is the mushy zone. If this zone has large width, the final microstructure will consist of large amount of distributed interdendritic shrinkage areas, as any feed metal from the riser will find it difficult to access many of these areas due to tortuous path involved. The width of the mushy zone is reduced as the cooling rate increases, as in metal mold castings, with consequent reduction in distributed shrinkage. When the mushy zone is absent or too small, the solidification is termed "skin-forming" and the feed metal from a properly designed riser will have good access to the solidifying areas, thus minimizing distributed shrinkage. The shrinkage under these conditions can be totally eliminated that the feed metal has access to the final solidifying area.

The application of Chroninov's rule, which states that the solidification time is proportional to the square of the volume-to-surface area of the casting and the riser or its modifications to account for the shape, will be helpful in this regard. The basic idea is to design the riser such that its solidification is more than that of the casting and its feeding distance is appropriate to reach the last solidifying zone of the casting. In long freezing range alloys solidifying in a mushy fashion, hot tear or hot crack can develop near above the solidus temperature when the network of solid crystals is unable to sustain any thermal stress gradients, particularly when the feed metal is unable to reach these locations. These cracks are usually sharp, capable of rapid propagation. Another important consideration in castings is the porosity caused by gas liberation during solidification. Gases like hydrogen are easily soluble in the liquid state but the solubility is substantially reduced in the solid state. This may result in pores of various sizes in the solid or even microcracks when there is significant resistance to the escape of the gases. It is therefore desirable to degas the liquid metal prior to pouring in the mold. A useful law in this context is Sievert's law which states that the solubility of a dissolved gas is proportional to the square root of its partial pressure. Using this law, degassing in the liquid state can be achieved by applying vacuum (difficult and expensive) or purging with an inert gas which serves the dual purpose of lowering the partial pressure of the dissolved gas and acting as a carrier for the escape of the dissolved gas, thus reducing the harmful effect of gas porosity in the solid.

As microstructure is the key to fine-tuning of the fracture toughness of castings, the influence of casting process factors on the microstructure must be well understood, if such fine-tuning is attempted. Needless to say, metallurgical knowledge such as phase diagram and the effect of non equilibrium cooling rate on it, nucleation and growth of the different phases in the microstructure, evolution of defects through impurities and interaction of the molten metal with melting atmosphere, the furnace lining, the mold, etc., will be very useful in this regard. Heat treatment can substantially affect the microstructure and therefore, knowledge of kinetics of solid state transformations is also important to understand the effect of the particular heat treatment on the microstructure.

Basics of Fracture Toughness Testing

Linear Elastic Fracture Mechanics (LEFM) Approach

Linear Elastic fraction mechanics approach may be defined as a method of analysis of fracture that can determine the stress required to unstable fracture in a component. The following assumptions are made in applying LEFM to predict failure in components.

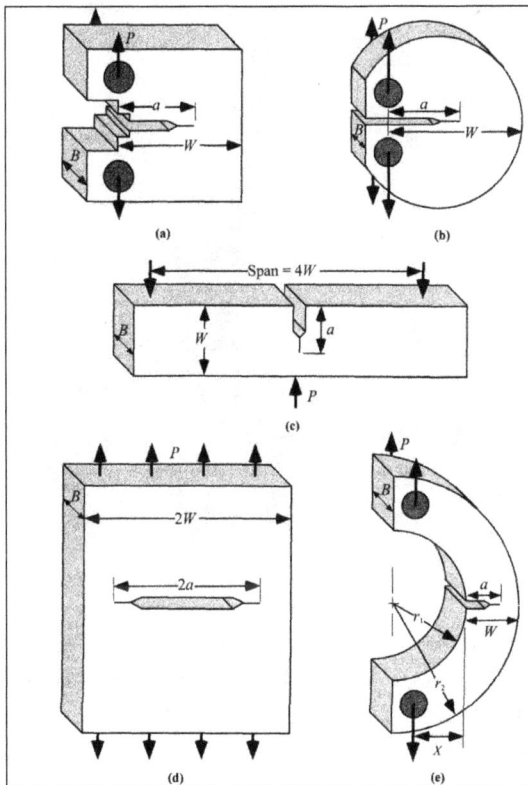

Standardized fracture mechanics test specimens: (a) compact tension (CT) specimen, (b) disk shaped compact tension specimen, (c) single-edge-notched bend (SEB) specimen, (d) middle tension (MT) specimen and (e) arc-shaped tension specimen.

- A sharp crack or flaw of similar nature already exists; the analysis deals with the propagation of the crack from the early stages.

- The material is linearly elastic.

- The material is isotropic.

- The size of the plastic zone near the crack tip is small compared to the dimensions of the crack.

- The analysis is applicable to near-tip region.

Figure below shows standardized test specimens recommended for LEFM testing. Each specimen has three important characteristic dimensions: the crack length (a), the specimen thickness (B) and the specimen width (W). In general, W=2B and a/w = 0.5 with some exceptions For brittle materials, a chevron-notch is milled in the crack slot to ensure that the crack runs orthogonal to the applied load.

A typical view of the test set up for fracture toughness testing.

In most cases fracture toughness tests are performed using either CT specimen or SEB specimen. The CT specimen is pin-loaded using special clevises. The standard span for SEB specimen is 4W maximum; the span can be reduced by moving the supporting rollers symmetrically inwards.

It is to be noted that the tip of the machined notch will be too smooth to conform to an "infinitely sharp" tip. As such, it is customary to introduce a sharp crack at the tip of the machined notch. Fatigue precracking is the most efficient method of introducing a sharp crack. Care must be taken to see that the following two conditions are met by the precracking procedure: the crack-tip radius at failure must be much larger than the initial radius of the precrack and, the plastic zone produced after precracking must be small compared to the plastic zone at fracture. This is particularly necessary for metal castings as many exhibit plasticity; a notable exception is flake graphite cast iron castings made in sand molds.

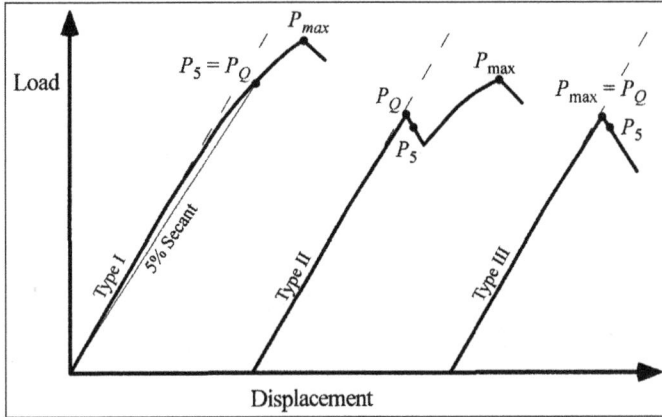

Type I, Type II or Type III behavior in LEFM test.

LEFM tests are conducted as per ASTM E 399. A typical test set up is shown in figure. All except the MT specimen noted in Figure are permitted to be used as per this standard. The ratio of 'a' as defined in each figure to the width W should be between 0.45 and 0.55. The load-displacement behavior that can be obtained in a LEFM test, depending on the material, can be one of three types as shown in the figure.

First a conditional stress intensity factor K_Q is determined from the particular curve obtained using:

$$K_Q = \frac{P_Q}{B\sqrt{W}} f\left(a/W\right)$$

Where, f (a/W) is a dimensionless factor of a/w.

The conditional stress intensity factor K_Q is the critical stress intensity factor if:

$$B, a \geq 2.5 \left(\frac{K_Q}{\sigma_{ys}}\right)^2$$

Where, σ_{ys} is the yield stress of the material.

If this is not the case, the result is invalid, most likely because of significant crack tip plasticity. This would imply that triaxial state of stress required to ensure plane strain condition at the crack tip is not achieved and any determined stress intensity factor at fracture as per ASTM E399 would be an overestimate of the resistance to crack growth. Use of such values in design would be dangerous. In such cases, an elastic-plastic fracture mechanics (EPFM) method must be employed to determine the specimen's resistance to the propagation of a sharp crack.

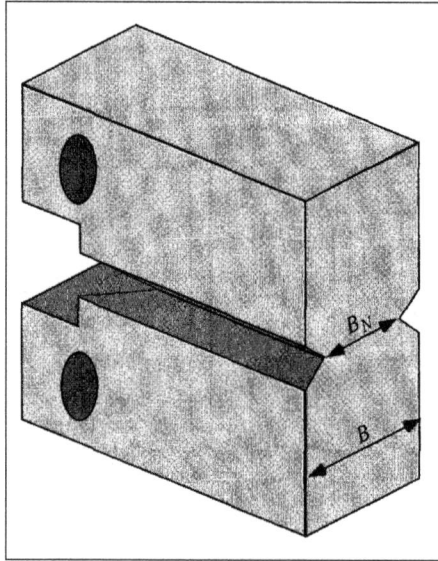

Side-grooved Fracture Toughness Test Specimen.

Elastic Plastic Fracture Mechanics (EPFM) Approach

Among the different methods available to determine the sharp crack growth resistance in specimens with significant plasticity at the crack tip (much less than what is required to cause total plastic collapse) the J-integral method and the Crack-tip Opening Displacement (CTOD) have been more widely adopted. The recent trend however, is to use the J-integral approach and only this method will be briefly described here. ASTM E 1820 gives two alternative methods: the basic procedure and the resistance curve procedure. The basic procedure normally requires multiple specimens, while the resistance curve test method requires that crack growth be monitored throughout the test. The main disadvantage of this method is the additional instrumentation and skill is required. Though this method has the advantage of using a single specimen, making of multiple specimens as nearly externally identical-looking castings is not a major problem; any inconsistent results among the different specimens will give an opportunity to see if the casting microstructure is properly controlled. Therefore only the basic test procedure will be considered here.

The Basic Test Procedure and J_{IC} Measurements

The ASTM standard that covers J-integral testing is E 1820. The first step is to generate a J resistance curve. To ensure that the crack front is straight the use of a side grooved specimen as shown in figure, is recommended.

A series of nominally identical specimens are loaded to various level and then unloaded The crack growth in each sample, which will be different is carefully marked by heat tinting or fatigue cracking after the test. The load-displacement curve for each sample is recorded. Each specimen broken open and the crack growth in each specimen is measured.

J is divided into elastic and plastic components, by using:

$$J = J_{el} + J_{pl}$$

$$J_{el} = \frac{K^2 (1-v^2)}{E}$$

$$K \frac{P}{\sqrt{BB_N W}} f(a/w)$$

$$J_{pl} = \frac{\eta A_{pl}}{B_N b_0}$$

η is a dimensionless quantity given by:

$$\eta = 2 + 0.522 (b_0 / W)$$

In equations $J_{el} = \dfrac{K^2 (1-v^2)}{E}$ and $K \dfrac{P}{\sqrt{BB_N W}} f(a/w)$ b_0 is the initial ligament length.

A_{pl} is the plastic energy absorbed by the specimen determined from figure below:

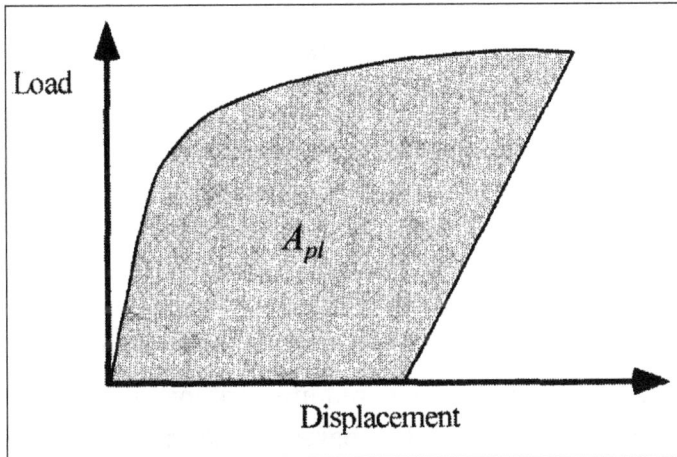

Plastic energy absorbed during J-integral test.

The J values obtained from equation $B, a \geq 2.5 \left(\dfrac{K_Q}{\sigma_{ys}} \right)^2$ are plotted against the crack extension Δa for each successive specimen to obtain J-R curve shown in figure.

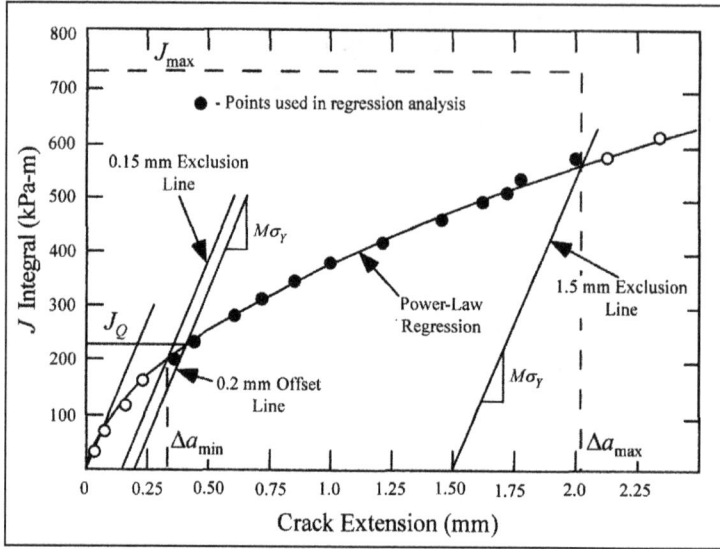

Determination of JQ as per ASTM E 1820.

M value in the figure is related to crack blunting and the default value is 2. As seen in the figure, the provisional critical value J_Q is obtained from the intersection of the J-R curve with the line M σ_Y where σ_Y is the flow stress given by the average of the tensile and yield stresses.

The provisional J_Q is taken as the critical value J_{Ic} if the condition:

$$B, b_0 \geq \frac{25 J_Q}{\sigma_Y}$$ is satisfied.

The equivalence between J_{IC} and K_{IC} is given by:

$$J_{IC} = \frac{(K_{IC})^2}{E}(1 - v^2)$$

Where, E is the elastic modulus and v is the Poisson's ratio.

If it is assumed that a steel sample has a yield strength of 350 MPa, tensile strength of 450 MPa and Young's modulus (E) of 207 GPa and fracture toughness of 200 MPa-$\sqrt{m}$, it can be shown that E 399 thickness requirement for validity is 0.816 m, while the E 1820 thickness requirement for validity, based on the equivalence shown in equation

$$J_{el} = \frac{K^2(1 - v^2)}{E}$$ is only 11 mm. The advantage of E1820 approach over E399 approach

in regard to valid specimen thickness requirement is thus obvious.

Fracture Toughness of Metal (Alloy) Castings

Aluminum Alloy Castings

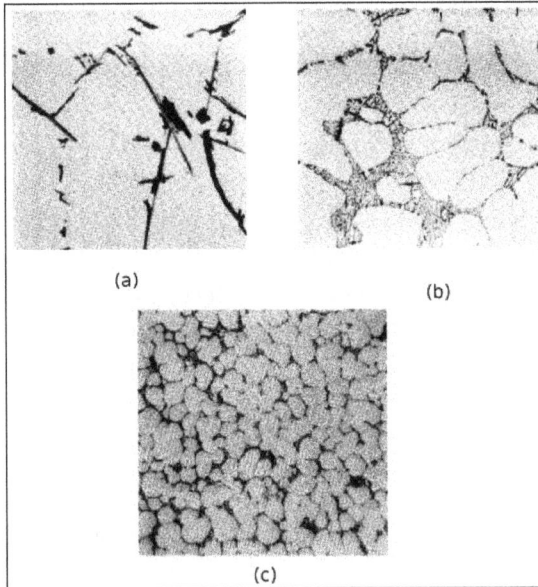

Microstructure of aluminum casting alloy 443 (Al-5%Si) (a) Alloy 443-F, as sand cast, (b) Alloy B443-F, as permanent mold cast, (c) Alloy C443-F, as die cast All were etched with 0.5% hydrofluoric acid and photographed at 500 X.

In recent times, the most widely studied nonferrous casting alloys for fracture behavior are aluminum casting alloys. Among them, aluminum silicon alloys have attracted the most attention as they are widely used because of good castability and high strength-to-weight ratio. The microstructure of aluminum silicon alloys can be significantly affected by changes in the process variables as typically shown below in figure for aluminum-5% silicon alloy. The figure shows the variation of microstructure with cooling rate.

Figure (a) refers to a sand casting where the cooling rate is the lowest among the three, sand cast, permanent mold cast and die cast. The dendrite cells are large, the silicon flakes (dark) are coarse and iron-silicon-aluminum intermetallics (light grey) are seen. The resistance to crack propagation will be the lowest with this type of microstructure. Figure refers (b) to a permanent mold casting where the cooling rate is higher than in a sand castings. It is seen that there is refinement in both primary aluminum and eutectic silicon as well as the intermetallics. The resistance to crack growth will be higher than in sand castings. Figure (c) shows the microstructure of a die casting of the alloy where high degree of refining of dendrite cells and eutectic silicon are seen. Other things being equal, the resistance to crack growth will be maximum in this type of microstructure. However other things will not be equal in general, the main factor being the yield strength of the casting. Thus crack tip plasticity will be high in the sand casting, intermediate in the permanent mold casting and lowest in die casting. Thus the fracture toughness increase in the casting will not be in direction proportion to the reduction in cooling rate. The factors favoring increase in fracture toughness would be decreased dendritic cell size and refinement of the covalent bonded silicon and mixed bonded intermetallics. The opposing factor would be reduced plasticity due to increase in yield strength, both due to primary cell and eutectic refinement.

Figure below refers to the microstructural variations brought by heat treating alloy 356-Al-7% Si-0.3%Mg (sand cast, constant cooling rate). Figure (a) refers the microstructure after artificial aging. The coarse dark platelets are silicon, black script is Mg2Si and the light scripts are intermetallics of iron-silicon-aluminum and iron-magnesium-silicon-iron. The dendrite cell is coarse. Figure (b) refers to alloy 356-F as sand cast, which is modified with 0.25% sodium. The cells are still coarse but the silicon particles are refined and in the form of interdendritic network. Figure (c) refers to alloy 356-T7 which is modified with 0.025% sodium, solution treated and stabilized. The microstructure shows rounded interdendritic silicon and iron-silicon-aluminum intermetallics. Here again the opposing factors discussed above will come into play but the plasticity in primary aluminum will be around the same and the dual causes for increase in yield strength as in the previous case will not be present.

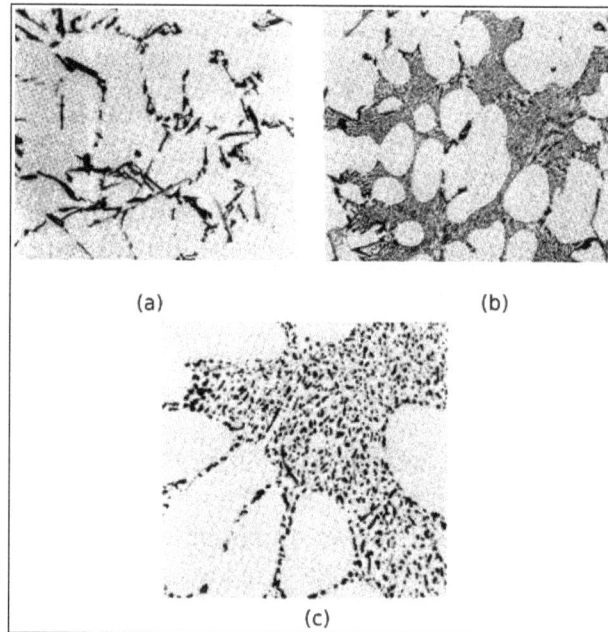

Microstructure of alloy 356 sand cast and heat-treated in different conditions (a) alloy 356-T51: sand cast, artificially aged, (b) alloy 356-F: as sand cast, modified with 0.025% sodium, (c) alloy 356 T7: sand cast, modified by sodium addition, solution treated and stabilized. All were etched with 0.5% hydrofluoric acid and photographed at 250 X.

Having noted these factors in affecting the fracture toughness, some recent papers on fracture toughness of aluminum alloys will now be examined.

Hafiz and Kobayashi studied the fracture toughness of a series of aluminum silicon eutectic alloy castings made in graphite and steel molds. The microstructure was varied by treating with different amounts of strontium. J-R curve obtained from multiple specimens was used to determine J_Q values. Extensive microstructural and SEM fractographic studies were made. They defined a ratio (λ/DE_{Si}) where λ is the silicon particle spacing and $(DE)_{Si}$ is the equivalent silicon particle diameter. They also defined the void growth parameter as $VGP = \sigma_y (\lambda/DE)_{Si}$. They found that the equation $J_Q = -9.94+0.38(VGP)$ is obtained in their samples, with J_Q varying in a straight line fashion from about 7 kJm⁻² to about 78 kJm⁻² when the VGP varied from 50 MPa to 200 MPa. Their main conclusion is that in eutectic Al-Si alloy castings, greater the refinement of eutectic silicon, higher will be the fracture toughness.

Kumai, et al, on the other hand focused on the dendrite arm spacing of alloy A356, (which is hypoeutectic) permanent mold and direct chill (semi continuous) cast tear test samples in their work. The area under the load-displacement curve was determined as the total energy and was divided into energy for initiation and propagation. It was found that in direct chill casting, both initiation and propagation energies increased with decrease in the dendrite arm spacing (DAS); decrease in DAS resulted only in increase of propagation energy in permanent mold casting. The fracture surface was perpendicular to the load in permanent mold castings while it was slanted in DC casting indicating higher energy absorption during the fracture process. This test could at best be qualitative in determining the fracture behavior.

Tirakiyoglu has examined the fracture toughness potential of cast Al-7%Si-Mg alloys. He has reported that based on Speidel's data a relationship of the form:

$$K_{IC}(\text{int}) = 37.50 - 0.058\,\sigma_{ys}.$$

It can be developed between the maximum (intrinsic) fracture toughness and yield strength of this alloy. However, as suggested by Staley there are several extrinsic factors such as porosity, oxides and inclusions that tend to lower the fracture toughness. If these extrinsic factors are eliminated the intrinsic fracture toughness can be higher, given by:

$$K_{IC}(\text{int}) = 50.0 - 0.073\,\sigma_{ys}.$$

Equation $J_{IC} = \dfrac{(K_{IC})^2}{E}(1-v^2)$ gives the potential maximum fracture toughness of the

Al-7%Si-Mg cast alloy in the absence of defects. A nice feature of this paper is the listing of dendrite arm spacing of different types of aluminum-silicon-magnesium alloy castings.

Tohgo and Oka have studied the influence of coarsening treatment on fracture toughness of aluminum-silicon-magnesium alloy castings. The alloy: Al-7%si-0.4%Mg was cast in permanent mold and solution treated for 6 hr at 803 K followed by aging for 6 hr at 433 K. One batch was tested in this condition while a second batch was further given a coarsening treatment at 808 K for 50 hr, 100 hr, 150 hr and 200 hr. J-R curves were constructed using 5 specimens and JQ values were determined. The fracture toughness increased to 27 MPam$^{1/2}$ after coarsening of silicon, as compared to 20.8 MPam$^{1/2}$ for uncoarsened sample. The authors attribute the improvement to the increased plastic deformation of α-Al owing to more uniform distribution of silicon particles, energy dissipation due to damage of silicon particles around a crack and the rough fracture path in the coarsened sample.

Kwon, et al have investigated the effect of microstructure on fracture toughness of rheo-cast and cast-forged A356-T6 alloy. Interdendritic silicon was observed in the microstructure of rheo-cast sample while there was alignment of cells in the cast-forged sample along with more uniform dispersion of silicon particles. Fractographs of fracture toughness specimens indicated cleavage type fracture in the rheo-cast sample while there was fibrous fracture in the cast-forged sample. As to be expected the fracture toughness of the rheo-cast sample was 20.6 MPam$^{1/2}$ while the cast-forged sample showed a fracture toughness of 24.6 MPam$^{1/2}$.

Alexopoulos and Tirayakioglu have determined the fracture toughness of A357 cast aluminum alloys with a few minor chemical modification. The raw stock for further machining required for studies was continuously cast with intent to keep porosity and inclusions at a minimum level. The continuous casting process is the patented SOPHIA process capable of providing cooling rates of up to 700 K/min. As compared to an investment cast sample, the dendrite arm spacing in the SOPHIA-cast sample would be lower by about 33%. The fracture toughness values, determined from CTOD measurements, ranged from about 18 MPam$^{1/2}$ to about 29 MPam$^{1/2}$, depending on the composition and the heat treatment. The higher value was obtained in the plain A357 cast by SOPHIA process and subjected to solution treatment for 22 hr at 538 C and aged for 20 hr at 155 C. The main aim of these authors was to establish correlation between tensile properties and fracture toughness and the major part of the paper deals with evaluation of tensile behavior under different conditions.

Lee, et al have investigated the effect of eutectic silicon particles on the fracture toughness of A356 alloy cast using three different methods: low pressure casting (LPC), casting-forging (CF) and squeeze casting (SC). They used ASTM E 399 procedure and as to be expected, got invalid fracture toughness results (sample thickness was 10 mm). They also conducted in-situ SEM studies on crack morphology, where plane stress was present. Thus only qualitative comparisons can be made on the influence of the three different casting processes on the fracture toughness. A notable observation is that significant shrinkage pores were present in LPC samples, while they disappeared in CF and SC samples, evidently due to the higher pressures applied. The eutectic cell size was the least in SC samples while it was similar in size in PC and CF samples. SEM fractographs from all the three samples showed fibrous fracture, with LPC samples showing the additional effect of stress concentration at the edges of shrinkage cavity. Though the SC sample had the most refined microstructure, the apparent fracture toughness was the lowest on account of reduced spacing between the eutectic silicon particles that apparently encouraged fracture initiation.

Tirakiyoglu and Campbell have analyzed the fracture toughness of Al-Cu-Mg-Ag (A201) alloy from data on premium quality castings. When molten metal is poured into a mold, the Reynolds number is invariably in the turbulent flow region to facilitate proper filling of the mold. In aluminum alloys, the surface oxide that forms as a result becomes folded into the bulk of the melt. These oxide "bifilms" have neutral buoyancy, unlike in say, steel castings and tend to travel with the melt into the mold cavity. As they do not bond with the liquid, the solidified casting will have the bifilms remaining as cracks due to the discontinuity. Also, the layer of air in the folded bifilms can grow into a pore or remain as a crack in the casting. The authors point out that in aluminum (and other drossing alloys) this is perhaps the most ignored defect as far as plans for elimination of defects are concerned. This extrinsic defect will result in the intrinsic fracture toughness not being attained. As per the authors, the intrinsic fracture toughness in A301 casting can be represented by:

$$K_{IC}\left\{In\left[1+\frac{\exp\left(-0.0032\,\sigma_{ys}\right)}{100}\right]\right\}^{3/2}\left(\frac{2kE'\,\sigma_{ys}}{3}\right)^{1/2}$$

Here,

$$E' = \frac{E}{1-v^2}$$

The intrinsic value of K_{IC} can exceed 45 MPa m$^{1/2}$ if the yield strength is around 350 MPa.

Steel Castings

Jackson has published a comprehensive paper on the fracture toughness of steel castings. He has considered that steel is susceptible to ductile-brittle transition and has reported the fracture toughness for lower shelf using LEFM and for the upper shelf using EPFM. While the LEFM method he used was the same as ASTM E399, use of CTOD was more in vogue in England at the time he wrote the paper and therefore either the critical CTOD (δ_c) or the equivalent J_{IC} have been reported in the paper, using the relation:

$$\delta_C = \frac{J_{IC}}{\sigma_{ys}}$$

Table: The fracture toughness, yield strength and chosen values of critical flaw size for three cases are shown in table.

Steel	K_{IC} (MPa m$^{1/2}$)	σ_{ys} (MPa)	Critical flaw size (mm) Surface Embedded
0.5%C, 1% Cr	46	480	3.7 4.4
1.5%Ni-Cr-Mo	86	740	5.4 6.2
1.5%Ni-Cr-Mo	104	1280	2.6 3.2

Steel 3 was vacuum melted while the other two were air melted, showing that a stronger steel has the disadvantage of lower critical flaw size (elliptical flaw, ratio of major-to-minor axis is 8-to-1).

One important point made by Jackson is that the chemical composition effects on fracture toughness may be masked by those of features such as shrinkage. Though it is known that increasing sulfur and phosphorus leads to decrease in fracture toughness, in the researcher's experiments, shrinkage masked this expected effect. Shrinkage encountered in the crack path may cause multiple crack fronts deviating from the main path resulting in increased fracture toughness to be observed; this overestimates the intrinsic fracture toughness and may cause problems when applied in design. The best remedy is therefore is to minimize shrinkage using proper feeding techniques.

As reported by Jackson, in the case of a 0.5%Mo, 0.33% V steel casting the lower shelf fracture toughness is about 55 MPa m$^{1/2}$ (temperature < 60 °C) while the upper shelf value increases to about 180 MPa m$^{1/2}$ (temperature > 110 °C). This behavior is inherent in BCC alloys like steel and should be considered in equipment where there is a wide difference between the cold start temperature and operational temperature. The problem then is to avoid brittle fracture during cold start and onset of plastic instability at normal operating temperatures.

Barnhurst and Gruzleski have investigated the fracture toughness of high purity cast carbon and low alloy steels. A notable feature of this work was that only blocks that were found to be radiographically sound were used for the preparation of fracture toughness specimens. The inclusion level in all the castings were low enough to classify them as extremely clean. The steel compositions were according to AISI/SAE 1030, 1527, 1536, 2330, 2517 for low carbon steels, 1040, 5140, 1552, 5046, 2345 for medium carbon steels and 1055, 5155, 3450, 52100 for high carbon steel. Other than carbon, each grade no other element or one alloying element, with impurities being kept to a minimum. All castings were austenitized in the range of 84 °C- 90 °C depending on the alloy, for 4 hr, oil quenched, held mostly at 650 C for 2 hr (with two exceptions: two samples directly air cooled from 90 °C soak, one sample held at 30 °C after oil quenching and then air cooled. The fracture mode in most castings was ductile, with only a few showing cleavage or mixed ductile/cleavage fracture. The K_{IC} values determined from J_{IC} ranged from 41.6 MPa m$^{1/2}$ (1.0 C, 1.61 Cr) to 247.8 MPa m$^{1/2}$ (0.25 C, 4.60 Ni). The conclusions drawn were that under carefully controlled composition, heat treatment, inclusion and impurity content, exceptional fracture toughness values at room temperature can be obtained, at the expense of tensile properties. The critical flaw sizes would exceed the section thickness of most designs. Under normal production conditions where attainment of such high purity is impractical, this study does provide the guidelines that the influence of alloying elements like nickel, chromium and manganese is relatively small at medium carbon levels and that heat treatment, additions of molybdenum and silicon may have significant influence on room temperature fracture toughness.

Chen, et al have studied random fracture toughness values of China Railway Grade B cast steel wheels using LEFM approach. The wheel was first stress relieved, and then the rim was quenched and tempered, while the hub was shot peened. K_Q values reported range from 50.52 MPa m$^{1/2}$ to 63.77 MPa m$^{1/2}$ in the wheel hub and, 60.70 MPa m$^{1/2}$ to 76.40 MPa m$^{1/2}$ in the wheel rim. Only the specimen thickness (~25 mm) has been indicated but the yield strength values have not been provided: it is therefore difficult to say whether these values are valid or not. Narrative description of the fracture surface using SEM indicates the predominance of cleavage with little evidence of fibrous rupture.

Kim, et al, have evaluated the fracture toughness of centrifugally cast high speed steel rolls. The carbon equivalent, defined as C + 1/3 Si was in the rang if 1.89 to 2.28 and the tungsten equivalent, defined as W + 2 M$_o$ was in the range of 9.82 to 13.34. Vanadium content was varied between 3.95 and 6.26 and the chromium content was kept constant in the range of 4.0-6.0. Precracking presented difficulties and therefore the authors used 30-50 µm machined notch. Tests were made otherwise as per ASTM 399. K_Q values were in the range of 21.4 MPa m$^{1/2}$ to 28.2 MP a m$^{1/2}$. They have concluded that the fracture toughness is determined by the total fraction of carbides, characteristics of the tempered martensitic matrix, distribution and fraction of intercellular carbides and fraction of cleavage and fibrous mode on the fracture surface. The best fracture toughness as obtained when a small amount of intercellular carbides was distributed in a relatively ductile matrix of lath martensite.

James and Mills have investigated the fracture toughness of two popular as cast stainless steels, CF8 and CF*M. Toughness tests were conducted at 24 °C, 371 °C, 427 °C and 482 °C using multiple specimen J-R curve method. Exceptionally high J_{IC} values, in the range of 1397 kJ/m^2 at 24 °C to 416 kJ/m^2 at 482 °C demonstrated that fracture control is not a concern in unirradiated condition. However, neutron irradiation reduces J_{IC} by an order of magnitude and therefore fracture control becomes essential.

Cast Iron

The metallurgy of cast iron is among the most complex of all alloys. Cast iron shows metastability anomaly. Under certain conditions of composition and cooling rate the eutectic formed upon solidification consists of austenite and graphite. Under certain other conditions an eutectic of austenite and iron carbide is formed. The former is known as graphitic cast iron, while the latter as white cast iron. In low sulfur and oxygen cast iron melts, if magnesium is added so that its residual amount is 0.05% or above (but not too high) the graphite formed will be nodular rather than the flake form found in untreated graphitic cast iron melts. In the latter the flake may be of undercooled type (Type D- when the sulfur content is low in sand or investment castings or with normal sulfur when the cooling rate corresponds to that in permanent molds); it will be in the interdendritic form, with branching). Under normal conditions found in commercial sand castings, the graphite will be a part of the eutectic cell formed with austenite, graphite having a loose "cabbage" shape with the interleaf region occupied by eutectic austenite. Adding a silicon-bearing inoculant will increase the number of eutectic cells in flake cast iron and nodule count in nodular (or, ductile) iron. The white cast iron forms graphite in the solid state when heat treated (and is called malleable iron), but the melt-formed graphite in the other two types of cast iron will be largely unaffected by any solid state transformation. In recent times another type called compacted graphite cast iron has been developed where the residual magnesium is lower than in ductile iron. All types of cast iron noted above are governed by eutectoid decomposition, which means that the matrix may consist of various combinations of ferrite and pearlite under near-equilibrium conditions. These irons are also affected by isothermal or continuous cooling transformations at nonequilibrium rates giving rise to bainitic or martensitic or tempered martensitic cast irons.

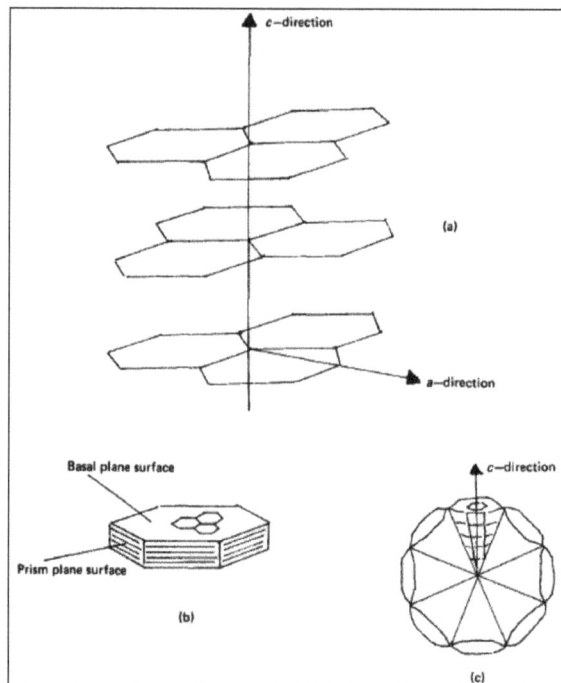

(a) Graphite laltice model, (b) Flake geometry, (c) Nodule geometry:
Extremities in the growth of graphite in the liquid.

The exact reasons for the formation of different types of graphite in cast iron have been a matter of debate for many years. The type of graphite found in commercial cast irons may have one or more of the following types: flake (Type A), undercooled (Type D), coral, compacted, nodular. A generalized view, based on the growth of graphite (in the liquid state) is presented in figure.

Graphite has a layered hexagonal lattice structure (a), with strong covalent bonds in the hexagonal chains, with the layers bonded by weak secondary bonds. The hexagonal plane is called the basal plane and the edge of the block formed by bonding of layers with weak bonds is called the prism plane. The basal planes tend to grow in the "a" direction and the prism planes in the "c" direction. When growth in the "a" direction is dominant, flake form is obtained, the thickness being determined by the growth rate and the graphite source; slower the growth rate and lower the number of eutectic cells, the thicker would be the flake. When "a" growth is suppressed and "c" growth is fully encouraged, nodular form results. Intermediate forms like Type D, coral or vermicular forms result when there is progressive resistance to the formation of Type A, or alternatively, decreasing encouragement to the nodular shape formation. It is to be realized that the exact reasons for these resistances or discouragements may be due to the interaction of fine-scale multiple activities, often at the atomic scale, related to both nucleation and growth. Thus, at this time one has to accept that these different forms of graphite, which curiously are relatively stable forms during the life of a component, do exist and there is need to understand, for instance, the details of how the crack propagation is affected by their interactions with the fine-scale features of the neighborhood of the crack.

The fracture toughness of graphitic cast iron is determined by the type of graphite, the type of matrix and the interaction between the graphite and the matrix. In view of numerous combinations possible, the fracture toughness could be expected to vary over a wide range: this is indeed the case. Once again it follows that to fine-tune the fracture toughness of cast iron the microstructural features should be analyzed and examined if corrective measures can be taken, consistent with cost-benefit analysis.

The fracture toughness of flake cast iron ranges from 11-19 MPa $m^{1/2}$. Whether these are valid results as per ASTM E399 is subject to the acceptance of the tensile strength instead of the yield strength for validity criterion, as flake cast iron has non linear elastic part in the stress-strain curve and 0.2% offset method can not be applied to determine the yield strength. Thus the above noted values may be cited by some as K_Q and by others as K_{IC}. In critical applications these low value force the assumption of a high factor of safety. A pertinent observation with respect to flake cast iron is that the ductile-brittle transition temperature is well above the room temperature and therefore the fracture toughness at normal or below-normal operating temperatures seems to be unaffected by the temperature.

Because of the steel-like mechanical behavior of nodular graphite cast iron, the fracture toughness of this iron has been vastly studied. The fracture toughness values range from about 25 MPa m$^{1/2}$ in an iron with yield strength of about 450 MPa to nearly 60 MPa m$^{1/2}$ in an iron of yield strength of 370 MPa. It is possible that the intrinsic fracture toughness of nodular iron would be higher if the inherent shrinkage, among the highest in cast irons, is reduced. A particular grade, D7003 (quenched and tempered) posses both good fracture toughness and high yield strength. Salzbrenner evaluated the fracture toughness of samples of different compositions, but adopted a constant heat treatment with intent to have a ferritic matrix. The heat treatment involved solutionizing at 90 °C for 4 hr, followed by slow furnace cool (at 10 °C/hr) to 70 °C and holding at this temperature for 24 hr followed by slow cooling. He followed EPFM approach and obtained fracture toughness values ranging from a high of 79 MPa m$^{1/2}$ (with small, well distributed nodules) to as low as 25 MPa m$^{1/2}$ in a sample with non-spherical nodules. The better fracture toughness of nodular iron in relation to flake cast iron is often attributed to the relatively smooth graphite-matrix interface in the former. This statement may however, be an oversimplification as there are factors such the diversion of the crack and ability to absorb energy in the interlayer regions of the nodule to be considered.

Doong, et al have investigated the influence of pearlite fraction on fracture toughness of nodular iron and their results show that when the pearlite fraction is 4% or 27% the fracture toughness shows a decreasing trend in the range of -75 °C to 75 °C, while the fracture toughness of samples with 67% and 97% pearlite show an increasing trend in the same temperature range. The nodularity in all these castings was 95% or better.

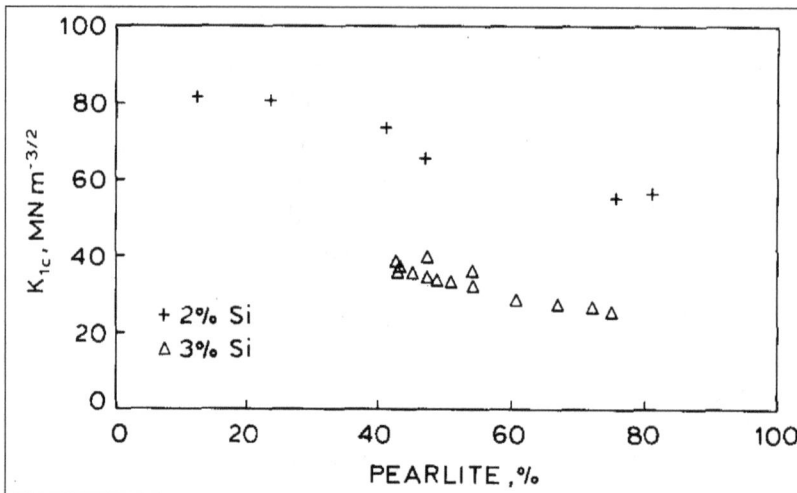

Effect of pearlite content on fracture toughness of permanent mold ductile iron.

Nodular iron castings are generally made in sand molds but the present authors investigated the fracture toughness of permanent mold-cast magnesium-treated iron. A hypereutectic composition with a high silicon percentage (3-3.4%) was used to avoid the formation of iron carbide in the as-cast state. The graphite consisted of overlapping nodules, possibly as a result of high thermal convection in permanent molds. It is also possible that inoculation was needed to provide more nucleation of graphite and reduce the possibility of overlapping, by rapid austenitic shell formation around the nodules.

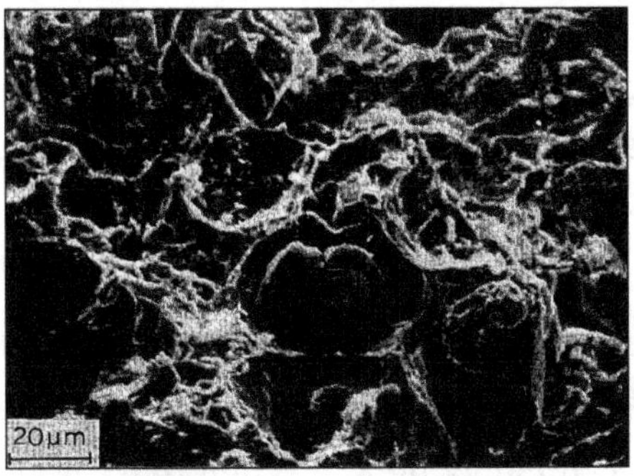

Fibrous fracture in the stable crack growth region of 2% Si casting.

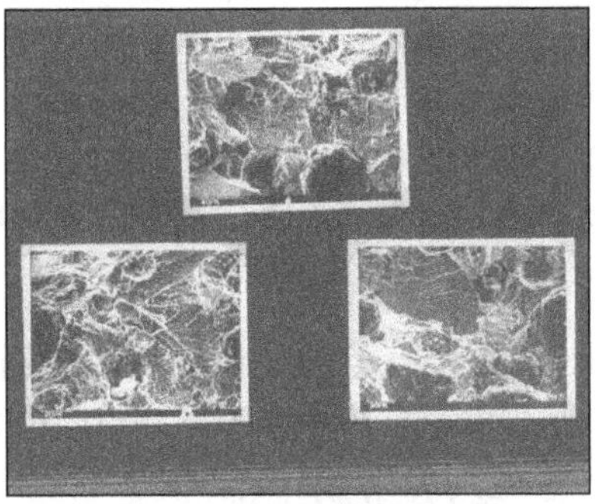

Transgranular cleavage in crack growth region of 3% Si casting.

Figure seen above shows the effect of pearlite content in two types of permanent mold ductile iron. The top curve refers to a melt with 2% silicon, which led to a chilled casting and was soaked at 90 °C and cooled at different rates to obtain different combinations of ferrite and pearlite in the matrix. The lower curve refers to a set of chill-fee castings, obtained by solidifying castings with 3% silicon. Different pearlite/ferrite combinations were produced by varying the casting thickness. It is seen that increase in silicon significantly lowers the fracture toughness. A possible reason is that on increasing the silicon level to 3%, the ductile-brittle transition temperature is raised to well above the room temperature. It is also possible that any residual stress present in the as-cast state is minimized in the heat treated state. In any case the differences in the modes of fracture in the two cases are clearly seen in figure.

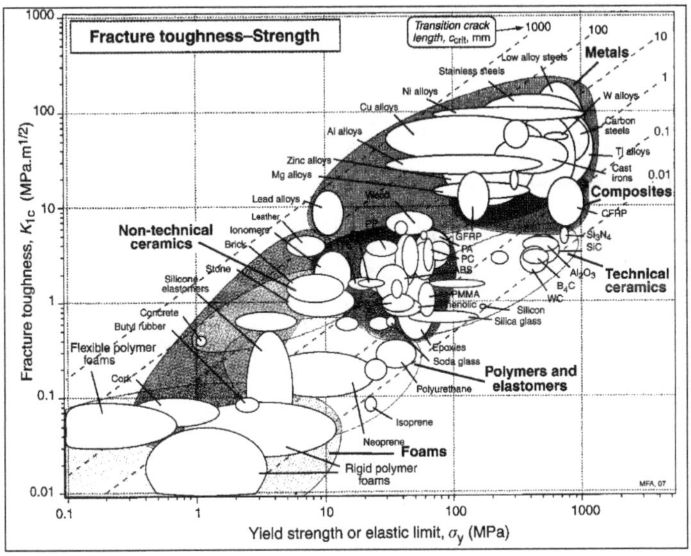

Fracture toughness, yield strength and transition crack length of materials.

A relatively new development in the field of ductile irons is Austempered Ductile Iron

(ADI) which is commercially available in different grades. The fracture toughness of ADI can be in the range of about 59-86 MPam$^{1/2}$ and therefore exceeds the fracture toughness of most other ductile iron grades, except Ni-resist. The fracture toughness values are best determined using EPFM. However, Lee, et al have used ASTM E399, which seems to be justified as the ratio 2.5 (K_{IC}/σ_{ys})^2 is below the test sample thickness of 25 mm; as the authors have not reported the yield strength, but only the Brinell hardness, the yield strength (MPa) is assumed to be 3.3 times the Brinell hardness, for the purpose of making this statement.

It is important to realize that both stress and crack size should be within limits for safe use of any casting. When the failure mode is brittle, the critical flaw size is given by equation $K_I = Y\sigma\sqrt{(\pi c)}$ when K_I reaches a critical value K_{IC}. When there is significant crack tip plasticity the transition from stable crack growth to unstable mode occurs at a length given by:

$$C_{crit} = \left(\frac{K_{IC}}{\sigma_{ys}}\right)^2 \frac{1}{\sqrt{\pi}}.$$

In the given figure is shown, a plot of fracture toughness versus yield strength with the transition crack length (mm), based on equation $K_{IC}\left\{In\left[1+\frac{\exp(-0.0032\,\sigma_{ys})}{100}\right]\right\}^{3/2}\left(\frac{2kE'\,\sigma_{ys}}{3}\right)^{1/2}$

of different values shown as parallel broken lines. All materials cut by a given transition crack line will have the same transition crack length. It would be a great benefit to the casting industry if similar charts are available only for casting alloys.

Corrosion-resistant Casting Alloys

The ability of a foundry to make good corrosion resistant castings depends on its ability to adhere closely to appropriate processing parameters. This capability is certainly within the reach of many foundries, but only if the economic drive exists to support their efforts. Five basic properties and capabilities must be developed to produce quality, corrosion resistant castings. These five factors include good alloy castability, good weldability, good corrosion resistance, the ability to accurately analyze the alloy, and the economic incentive to produce parts.

Good Alloy Castability

To develop a good casting alloy from a wrought alloy usually requires a modification of the wrought alloy chemistry. It might seem unusual to discuss this variable first rather than good corrosion resistance, but it does neither the foundry nor the ultimate user little good to develop a material of outstanding corrosion resistance if it cannot be produced in a consistent manner as a cast shape.

Nevertheless, many alloys are commercially cast even though they have less than ideal castability. For example, CN7M or UNS N08007, commonly referred to as Alloy 20, has exceptional corrosion resistance. In sulfuric acid applications, few materials can match its broad chemical resistance. It was developed as a cast alloy in the late 1930's, and was quickly introduced into a wide range of parts. However, as castings of this alloy entered service, it quickly became evident that the material had poor resistance to cracking and tearing. Twenty to thirty years passed before the foundry industry fully understood why this material was so susceptible to cracking and tearing, and determined what needed to be done to avoid these problems.

A second example is CD4MCuN or UNS J93372, one of the earliest cast duplex stainless alloys, developed in the early 1960's. In addition to good corrosion resistance, it has exceptional strength. However, early castings of this alloy were inferior. They exhibited delayed brittle fractures, and were very difficult to produce with consistent dimensional properties. As with CN7M, the foundry industry eventually learned how to resolve these problems, but only after 15 to 20 years of on-and-off difficulties.

Similar problems still exist today for the newer duplex and high moly austenitic stainless steels. These stainless steels have been alloyed to the practical limits of current foundry technology, and as a result many foundries are experiencing difficulty in producing castings with acceptable quality and with the expected corrosion resistance. Therefore, it is important when purchasing these newer stainless steels and higher alloys in general, that purchasers choose foundries that are known to have the expertise to cast them successfully the first time.

Ease of Weldability

Weldability is also absolutely essential to casting manufacture. Although some casting techniques minimize the need for weld repairs in some configurations, many of the castings that end up in pumps and valves are either weld-repaired or weld-fabricated at some stage of their manufacture. To have good weldability, the alloy must have good resistance to cracking and tearing during welding, and must be able to be welded with filler materials that match the mechanical and corrosion properties of the base metal.

Equally important for most commercial applications is the ability to qualify weld procedures to ASME Section IX requirements. Because shielded metal arc (SMAW) and gas tungsten arc-welding (GTAW) are the most common processes, it is critical that welding procedures be developed in at least one and preferably both of these practices.

Good resistance to chemical attack is a given requirement for corrosion-resistant castings. In general, the cast alloy must exhibit corrosion resistance comparable to the wrought product that it is complementing. To develop this corrosion resistance, the cast producer must understand the interaction between chemical composition, thermal history, and corrosion resistance. Such interactions will likely be different for the cast alloy than the wrought alloy. For example, a wrought alloy was developed that exhibited excellent resistance to 98% nitric acid. This wrought alloy was optimized as a wholly austenitic alloy with a nominal 4% silicon level.

To produce a cast alloy with equivalent corrosion resistance, the silicon level had to be increased to 5% and the chemistry balance had to be altered so that the alloy contained several percent ferrite in its microstructure. If the wrought chemistry had simply been duplicated, the cast material would have had inferior corrosion resistance and would have been extremely susceptible to cracking and tearing during casting and welding.

Accurate Chemical Analysis

Good castability, weldability, and corrosion resistance all depend on control of chemistry in a fairly narrow range. Therefore, the ability to accurately analyze alloy chemistry is critical. To generate reliable chemical analyses, reference standards must be available, preferably eight to ten at a minimum. These standards must bracket the expected control limits for the alloy for each element considered critical. The standards should include not only the major alloying additions such as chromium, nickel, copper, and molybdenum, but also trace elements such as sulfur, phosphorus, and carbon.

Attempts to rely on a single reference standard to analyze chemical composition can result in significant errors if the chemistry deviates only a slight amount from that of the single reference standard. Too many interactions are possible between alloying elements in corrosion-resistant castings to depend on a single sample to control alloy chemistry in a narrow range.

Economic Viability

The final factor and probably the greatest incentive needed to produce a quality corrosion resistant casting is economic viability. As noted earlier, considerable up-front effort must be made to produce a quality corrosion-resistant casting. To support this up-front effort, some payback must be possible to the casting producer. A single order for a few castings might satisfy the customer, but the producer cannot come close to covering his up-front costs with a single order. Repeat business is needed to justify the investment in engineering and sampling time. It also will enable the foundry to recycle melt stock.

For every pound of metal poured, only 0.4 to 0.6 pounds of usable casting are produced. The balance of the material must either be sold for scrap or recycled back into more castings. For a single order, the volume generally does not permit material recycling during this limited window. The foundry must either raise the price to compensate for excess material, or recycle it into other cast alloys.

However, recycling is not always a straightforward process. Corrosion-resistant cast alloys contain additives such as copper or tungsten, which enhance corrosion resistance of the alloy to which they are deliberately added, but they can have disastrous effects on the ability to recycle excess material. For example, copper is added to cobalt-base alloys to enhance corrosion resistance. Several corrosion resistant cast alloys contain this element, yet the vast majority of cobalt alloys produced do not contain copper, and many have fairly low permissible residual levels. If the foundry produces a corrosion resistant cobalt alloy with copper added to enhance corrosion resistance, it must either be able to count on repeat business in this alloy to enable recycling, or lose $10 or more per pound selling in a scrap market where no buyers exist.

After corrosion-resistant wrought alloys have been designed, complementary casting alloys must be developed to produce parts such as joints, pumps, and valves. Foundries must typically make large investments of time and money to acquire accurate data for casting alloy development.

Cast Alloys based on Wrought

Most new corrosion-resistant casting alloys are developed by wrought producers, many times in conjunction with the ultimate customer. This is certainly the appropriate starting place, since the bulk of the product sold will be in the form of plate, sheet, tube, and bar. Research and evaluation of the wrought alloy may proceed for several months up to several years, and certainly costs the wrought producer a considerable investment in time and resources. The end result is a proprietary, if not patented, alloy with chemistry optimized to produce high quality, low-cost wrought products.

When the new alloy becomes successful in corrosion applications, an immediate need is created for equivalent cast products to produce parts such as pumps and valves. Basic data available to produce these castings include general chemistry limits for wrought alloys; basic properties of the new wrought grade, including corrosion resistance and mechanical data; limited data on weldability of the wrought grade; and a proprietary alloy name.

However, little data is available that is considered essential to the manufacture of quality castings. The foundry or foundries asked to produce cast products in these new alloys must develop much of the same data that the wrought manufacturers needed. Such data includes basic alloy chemistry to make the alloy castable, maintain levels of corrosion resistance, and meet some reasonable mechanical property requirements. Weldability must also be established, and weld procedures developed to meet ASME requirements.

Standards required for chemical analysis must be custom made, since at best one or two standards might be available commercially. The shrinkage rate for the material as it solidifies and cools to ambient temperature must be determined, to decide if near net shape dimensional requirements may demand new or modified tooling. Although thermal requirements for wrought products could be known, no data will be available about whether the cast product will respond to these same thermal treatments in an equivalent manner. These and other questions lead to a series of investments for the cast producer that must be made if a high quality cast part is to be produced.

Advanced Technology

Several new technological developments can allow foundries to shorten their development time. Such technologies as stereolithography, solidification modeling, and computational systems design can shorten the development time and reduce the cost of developing new alloys significantly over traditional empirical methods.

Stereolithography is a means of rapid prototyping polymeric replicas that can then serve as an investment casting pattern to produce the final metallic part. Rapid prototyping is also suitable for making tooling for investment castings in a much shorter time, so that multiple samples can be made without having to make each directly from the rapid prototyping equipment. This technology has greatly improved a foundry's ability to produce and test new casting designs in a much shorter time frame.

Solidification modeling programs allow a foundry to take a CAD drawing of a part and, through modeling of shrinkage characteristics, determine the optimum gating and risering system for that part. Solidification modeling allows foundries to produce castings with the minimum number of gates and risers, yet produce a sound, quality part, usually on the first attempt, and at a lower cost.

Material development also benefits greatly from advances in computer technology. Better computers and computerized instrumentation enable research on a molecular and even atomic level. This nanotechnology, and computerized instruments such as the scanning probe microscope, allow observation and manipulation of atoms and molecules to make new or existing materials with enhanced properties. Today, this technology is already moving from the universities into commercial reality.

Computational systems design can shorten the development time and reduce the cost of developing new materials significantly over the traditional empirical method. This can allow for the development of custom materials for low volume applications for which empirical methods would have been too costly and impractical. Soon, foundries may be able to use this technology to shorten the development time when designing complementary cast alloys from wrought alloys.

Residual Stresses in Casting

The term residual stresses means: "all those stresses contained inside a body when the latter is in conditions of balance with the surrounding environment", and therefore without having any external evidence, to the extent that they are defined also as "internal stresses".

"Internal" or "residual", no matter how we define them, in a metal material such stresses always derive from some inhomogeneity conditions inside the material, inhomogeneity that in castings is generally coupled with the fact that the casting cooling does not occur simultaneously in all points, since the inner surface cools faster than the material core, and the zones with subtle walls before the more massive parts.

In general terms "the residual stresses are generated inside the component when inhomogeneity elements are present in the metallurgical or material process characteristics".

ERRORI DI FORMA O DIMENSIONE				
F100 : errori di forma, dimensioni corrette				
	F110 : non correttezza di tutte le dimensioni			
		F111 : tutte le dimensioni proporzionalmente non conformi	tolleranza di ritiro non adeguata	
	F120 : non correttezza di alcune dimensioni			
		F121 : eccessiva distanza tra le etremità	impedimento alla contrazione	
		F122 : errore su alcune dimensioni	contrazione irregolare	
	F230 : deformazioni rispetto alla forma corretta			
		F233 : getto deformato rispetto al disegno: modello e stampo conformi al disegno	distorsione del getto	Pattern / Mold / Casting
		F234 : getto deformato rispetto al disegno dopo immagazzinamento, distensione, lavorazione	svergolamento del getto	

Classification of geometrical-dimensional defects imputable to internal stresses according to the classification International Committee of Foundry Technical Associations (ICFTA).

In the case of castings, apart from macroscopic errors leading to have homogeneity differences of the material among the various zones of the part, the main source of residual stresses is undoubtedly the cooling process or, better, the non-simultaneous cooling among the various zones of the same component.

Reminding in fact that the specific volume of a metal material is directly proportional to temperature, it results that when the cooling of the external casting part occurs, the internal part is constituted by high-temperature material. When the cooling extends to the material core, too, this will not shrink freely because constrained by the external solidified surface: therefore, we will have the external part stressed by compression by the core material, which in its turn will be stressed by traction by the external part.

Analogue considerations are valid if the casting features parts with different thickness, with parts with subtle stabilized walls before the parts with bigger section, and then the establishing of mutual traction or even of deformations.

We can therefore state that the stress state in a casting with subtle walls has at least two components, one that we may define "vertical" induced by the interactions between the material on the surface and the underlying material, and a "geometrical" one induced by the constraints constituted by the surrounding material.

We should just consider that in the international classification of casting defects, as standardized by ICFTA – International Committee of Foundry Technical Associations), 7 macro categories are provided, which we remind here as follows:

- Bumps,

- Cavities,

- Discontinuities,

- Surface imperfections,

- Incomplete casting,

- Dimensional error,

- Inclusions or structural anomalies.

Inside discontinuity defects (G category), that is to say breakages detected at the extraction from the die, it is expressly provided the G200 sub-category "Discontinuities caused by internal tension and restraints to contraction (cracks and tears)".

Whereas in dimensional error defects (category F) there are a good 4 non-conformity typologies ascribable to residual stresses. As we have often repeated, however, the main danger of residual stresses is not that they lead to the breakage or deformation of the component at the extraction from the die but instead that they remain "hidden" inside the component, drastically decreasing the resistance in operation – for this reason, then, it becomes essential the capability of measuring residual stresses in castings, at first as process setup instrument and afterwards as tool of product quality control.

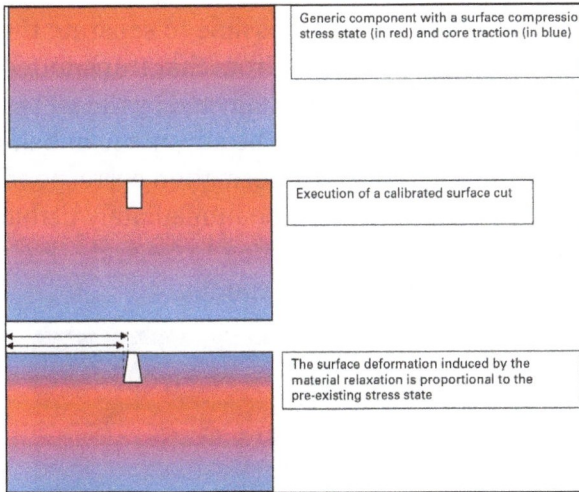

Generic component with a surface compression stress state (in red) and core traction (in blue)

Execution of a calibrated surface cut

The surface deformation induced by the material relaxation is proportional to the pre-existing stress state

Exemplification of relaxation by stock removal.

Measurement of Residual Stresses: Cutting, Ring Core and Hole Drill

Essentially, the measurement of the residual stresses contained in castings occurs by exploiting the measurement by means of strain gages, from the component deformation, following the removal of a controlled quantity of material or as consequence of the breakage on an internal constraint.

The execution modalities of such operations make the measurement more or less destructive for the component: the most destructive methodologies are generally used for the qualification and the development of processes while less invasive methodologies are used for the quality control of castings anyway intended for use.

The casting of an engine block: with its complex geometry and its alternation of walls with different thickness, it is a typical applicative example for the cutting technique.

In the application of the cutting technique, the method provides for the application of the strain gage on the zone to be investigated and for the "separation" of that zone from the adjacent material by means of a series of cuts (hence the method denomination).

Depending on the positioning of cuts, it is possible to separate the induced stress status, for instance by surrounding constraints, from that transmitted "on the surface" by more internal stresses. This method is clearly affected by the fact of being fully destructive for the analysed component but it offers the advantage of being easily applied and interpreted (considering that the released deformation corresponds to the entire present stress status, the pre-existing stress can be immediately deducted from the elastic constant of the material), as well as of permitting to separate the internal contribution of the material from that of surrounding constraints.

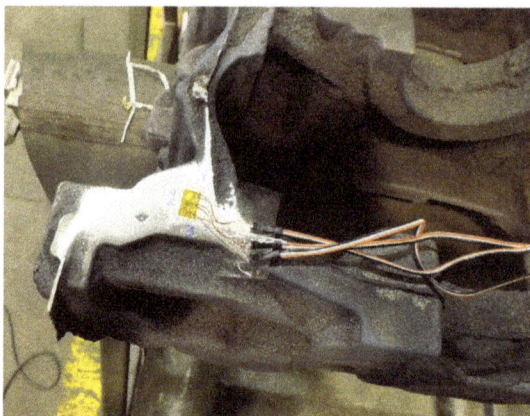

The positioning of the strain gage in the zone to be investigated.

Highlighted by the red arrows, the three cuts through which they have provided for eliminating from the monitored zone the stress effects induced by the surrounding parts: starting from the deformation detected by the strain gage (here protected by a specific protection), it is possible to trace back the stress condition before the cut execution.

In this case, the progressive execution of different cuts has allowed determining separately the stress state induced by the adjacent parts from the specific one of the material in the considered point (ex.by heat treatment).

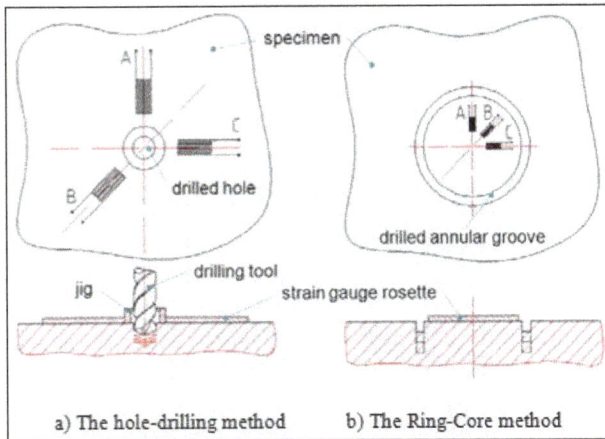

a) The hole-drilling method   b) The Ring-Core method

The different approach of the hole-drill and ring-core technique: in the first, it is made a hole of very small sizes (generally 2mm of depth and 1.6mm of diameter) and the deformation is detected outside the hole, in the second the material removal occurs outside the hole, carrying out a sort of "cutting" in miniature.

In the case of ring-core and hole-drill techniques, on the contrary, the stock removal is much more controlled and much smaller in quantity (even minimal in the case of the hole drill). In simple words, we can affirm that in the ring-core technique, the material is removed around the strain gage, which once more directly measures the complete material relaxation, while in the hole-drill technique the removed material is reduced to a very small hole of 2mm of depth and 1.6mm of diameter, and the deformation is measured radially outside that hole. In this case the material relaxation is only partial and the reconstruction of the overall stress is accomplished by an integral calculation algorithm, which considers the deformation gradually released hand in hand with the increase of the hole penetration depth: a measurement whose procedure requires specific strain gages, a highly specialized drilling equipment and relatively complex calculation algorithms.In spite of this, also thanks to the minimal destructivity for the component, of the two last techniques undoubtedly the hole-drill is more used, to the extent of being the subject of a specific international regulation (ASTM E-837 Standard Test Method for Determining Residual Stresses by the Hole-Drilling StrainGage Method).

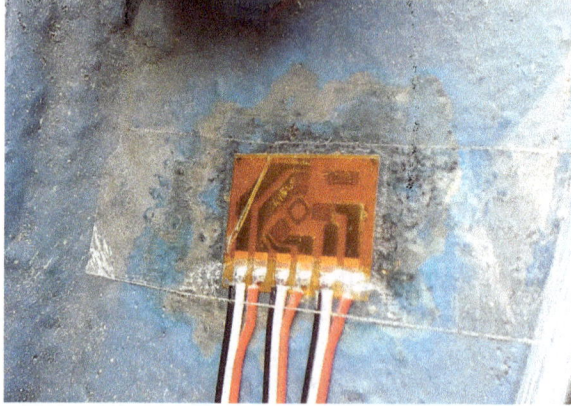

A specific strain gage for measuring the residual stresses by means of hole-drill technique with highlighted tree measuring grids and the localization provided for the hole.

The need of positioning the drill cutter with absolute precision and the need of executing the drilling at controlled depth requires the use of specific drilling equipment.

Residual stresses are a very dangerous phenomenon for castings, not so much for the possibility that they cause the breakage or the deformation of the component during the extraction from the die and the successive machining but instead because they can induce in the component a stress status absolutely free from external evidences but able to decrease dramatically the resistance in operation of the component. Fortunately, they have developed measuring techniques able to quantify with excellent accuracy the stress states in castings, exploiting the principle that a material removal results in a deformation of adjacent zones.

Properties of Moulding Sand

Permeability

The passage of gaseous materials, water and steam vapour through the moulding sand is related to porosity or in other words permeability.

The permeability of sand depends upon the following factors:

- Size of the grain (varying over the wide range of 50 microns to 3360 microns).

- Shape of the grain (round, angular, sub-angular or compound), the round shape being more favourable for porosity.

- Compactness density has also a bearing on the permeability.

- Moisture content in the moulding sand affects permeability since the excess moisture tends to collect in the interstices.

- Bonding content also affects the porosity of the moulding sand through the interstitial structure.

The sand used for casting must be porous enough, so as to allow the gaseous material, water and steam vapours to escape freely when the molten metal is poured into the mould. Molten metal contains some dissolved gases, which are evolved on solidification.

Further, molten metal in contact with moist sand forms steam and vapour which must find passage to escape completely. Insufficient porosity of moulding sand leads to casting defects such as gas holes and pores. The moulder has some control over permeability; hard ramming lowers the permeability, but this is relieved by liberal venting.

Cohesiveness

The ability of sand particles to stick together is termed as cohesiveness or the strength of the moulding sand.

Strength or cohesiveness of sand depends upon the following factors:

- Grain size and shape, which affect the strength characteristics to a considerable extent.

- Mixture of various-size grains.

- Bonding material or bond content and its distribution. Bond strength is determined by the alumina (clay) content. Clay should be present as a thin, tenacious film on each grain of sand. Sharp sands having smooth oval grains are not easily bonded and clay helps in bonding.

- Moisture content—a major factor that affects strength of sand.

Moulding sand must have a good strength, otherwise it may lead to collapse of the mould or its partial destruction during conveying, turning over or closing. Cohesion must be retained, when the molten metal enters the mould (bond strength) and sand should not be washed away during pouring.

Adhesiveness

The sand particles must be capable of sticking to the other bodies, particularly to the moulding box and it is only due to this property that sand mass is held in the moulding box properly and does not fall when the mould is moved. At the same time, the sand must not stick to the casting and strip off easily, leaving a clean surface.

Plasticity

It refers to the condition of acquiring predetermined shape under pressure and to retain it, when the pressure is removed. In order to have a good impression of the pattern in the mould, moulding sand must have good plasticity. Generally, fine grained sand has better plasticity. It depends on the content of clay, which absorbs moisture, when sand is dampened.

Refractoriness

It is the ability of the silica sand to withstand high heat without breaking down or fusing. Sand with poor refractoriness may burn at high temperature. The fusion point of moulding sand can be increased by removing the impurities, particularly metallic-oxides. Sand also should not form glassy mass which hampers stripping.

Chemical Resistivity

Moulding sand should not chemically react or combine with molten metal so that it can be used again and again.

Binding Property

Binder allows sand to flow to take up pattern shape. It must not be so strong that break out becomes difficult, nor should it be so weak that it allows surface skin of casting to break.

Flowability

This is similar to plasticity. It is the ability of sand to take up the desired shape. Sand must be able to transmit the blows throughout during ramming.

Green Strength

It depends on:

- Grain size,

- Shape and distribution of sand grains,

- Type and amount of clay or other binder, and

- Moisture content.

Dry compressive strength of a moulding-sand mixture increases as moisture is added, until the sand is too wet to be workable.

References

- Cast-iron-mechanical-properties: iron-foundry.com, Retrieved 3 June, 2019

- Evaluating-the-tensile-properties-of-aluminum-foundry-alloys-through-reference-castings-a-review: researchgate.net, Retrieved 19 May, 2019

- Science-and-technology-of-casting-processes/fracture-toughness-of-metal-castings, books: intechopen.com, Retrieved 28 April, 2019

- How-to-determine-residual-stresses: metalworkingworldmagazine.com, Retrieved 21 July, 2019

- Properties-of-moulding-sand-materials-casting-metallurgy, casting, metallurgy: engineeringenotes.com, Retrieved 15 January, 2019

Casting Alloys

There are a wide variety of alloys which are chosen for their diverse physical and mechanical properties. A few of the categories of alloys used for casting are aluminum-silicon alloys, zinc casting alloys and magnesium alloys. This chapter discusses in detail the casting processes related to these alloys.

The wide range of casting alloys helps in selecting the most suitable and cost effective material that meets the requirements of a particular application. Each of these casting alloy has its own physical and mechanical properties. They also have their own casting characteristics, like:

- Weldability,
- Machinability,
- Corrosion resistance,
- Heat treatment properties.

The continuous research on alloying elements often results in the invention and development of more powerful and suitable casting alloys to meet the demands of industry applications. Selecting the proper casting method and the most suitable alloy are two major things that help in achieving optimum cost level for a specific.

There are two major categories for casting alloys:

Ferrous Alloys

Ferrous alloys are iron based alloys that has extensive use in wide range of industries because of its flexibility to meet strength, toughness, and impact of diverse industrial applications.

Steel

- The properties of steels, determined by dispersion strengthening, depend on the amount, size, shape, and distribution of cementite (Fe_3C).

- These factors are controlled by alloying and heat treatment.

Surface Treatments of Steel

- Surface Heat Treatment: The surface is quickly heated, quenched, and then tempered.

- Carburizing: Diffusion of carbon into the surface to increase the carbon at the surface.

- Nitriding: Similar to Carburizing but nitrogen (N) is substituted for carbon.

Stainless Steel

- Ferritic Stainless Steels (BCC): Up to 30% Cr and less than 0.12% C. Good corrosion resistance.

- Martensitic Stainless Steels: Cr < 17%. Heat treatable with the ability to form martensite among other phases.

- Austenitic Stainless Steels (FCC): Ni is an austenite stabilizing element.

Cast Iron

- Grey Cast Iron: Interconnected graphite flakes in pearlite matrix. Good in vibration damping.

- White Cast Iron: Used for their high hardness and wear resistance. Martensite can be formed.

- Malleable Cast Iron: A heat treated unalloyed 3% carbon white iron.

- Nodular Cast Iron: The addition of magnesium (Mg) causes spheroidal graphite to grow.

Nonferrous Alloys

Nonferrous alloys contain no iron at all and normally are costlier than ferrous alloy. The copper alloys are the largest product group among these alloys.

Brasses and bronzes are the most popular copper alloys:

- Copper is alloyed with zinc to make brass. Most kinds of brass are easily shaped and have a pleasing appearance.

- Copper is alloyed mostly with tin to make bronze.

Alloys used in Die Casting

Die casting molds are usually constructed from hardened steel and they are often the most expensive component in a die casting machine. These molds can handle a range of different alloy families with varying results, but die casting is generally most effective on metals with low fusing temperatures. For this reason, the common die casting alloys fall into a handful of categories based on their composition and material properties.

Zinc Alloys

Zinc-based materials are relatively easy to die cast, and respond well to the die molding process. These materials are comprised of multiple metals in specific ratios. For example, a typical zinc-based die casting workpiece consists of 86 percent zinc, 4 to 7 percent copper, and 7 to 10 percent tin. Slightly higher proportions of tin make the workpiece more flexible, while increased copper levels improve rigidity. Zinc alloys have a melting point in the range of 700 to 800 degreees Fahrenheit.

Zinc die castings are often used in place of cast iron or brass, but tend to have lower tensile strength than their sturdier counterparts. Unless it is specially reinforced during the alloying process, zinc-based material cannot exceed approximately 17,000 pounds per square inch of force. As a result, die cast zinc products are generally not used in applications involving high mechanical loads. Zinc castings can also be corroded by alkaline substances or salt-water, and are often plated to preserve their luster despite atmospheric conditions.

Tin Alloys

Tin Alloys are most often used in applications requiring corrosion resistance, such as those involving the food industry or internal and external bearings. While the proportion of metals in these alloys can vary widely, a typical tin alloy consists of 90 percent tin, 6 percent antimony, and 4 percent copper, which is added to strengthen the material's durability. Tin alloy die castings generally weigh under ten pounds and rarely exceed 1/32 of an inch in thickness. They are valued for their resistance to alkaline, acids, and water, but feature a comparatively low tensile strength rating of below 8,000 pounds per square inch.

Bronze and Brass Alloys

Most bronze and brass materials can be die cast as effectively as zinc-based alloys, although small holes can only be drilled into the workpiece after casting, rather than during the casting process. Bronze and brass are commonly used to create washers, camshaft components, and decorative products (due to their distinctive coloring and potential for surface finishes). A typical brass alloy consists of 60 percent copper, 40 percent zinc, and 2 percent aluminum, but there are many variations on this mixture. Die casting bronze and brass is capable of yielding products with a durable surface and highly accurate interior specifications.

Some brasses have difficulty tolerating shrinkage from high temperature processes, but despite these challenges, most of these alloys can be used for products weighing up to fifteen pounds and with thicknesses at or under 1/32 of an inch. They are generally suitable for applications requiring tensile strength of less than 8000 pounds per square inch.

Aluminum Alloys

Die cast aluminum alloys are often found in automobile parts and gears, and have been used to create surgical instruments in the past. They are generally stronger and lighter than most zinc-based materials, but tend to be more expensive to create. Using aluminum alloys can reduce the need for finishing treatments, such as plating, and a common grade is composed of 92 percent aluminum mixed with 8 percent copper. Magnesium may be added to this alloy to improve its tensile strength from around 21,000 pounds per square inch to approximately 32,000 per square inch, while nickel can be included to increase rigidity and provide a higher surface finish. The melting point for an aluminum alloy is around 1150 degrees Fahrenheit.

Lead Alloys

Like tin alloys, lead-based materials tend to be used for their corrosion resistance and in applications requiring no more than 8000 pounds of tensile strength per square inch. Common applications include fire-safety equipment, bearings, and various decorative metal goods. They are relatively inexpensive for producing castings under 15 pounds, but lead alloys cannot be used for products that will be in contact with food. A typical lead alloy might be 90 percent lead and 10 percent antimony, with tin being a common addition as well. The melting point is usually around 600 degrees Fahrenheit, and product thickness rarely exceeds 1/32 of an inch.

Cast Iron Components

Cast irons typically contain 2-4 wt% of carbon with a high silicon concentrations and a greater concentration of impurities than steels. The carbon equivalent (CE) of a cast iron helps to distinguish the grey irons which cool into a microstructure containing graphite and and the white irons where the carbon is present mainly as cementite. The carbon equivalent is defined as:

$$CE\left(wt\%\right) = C + \frac{Si + P}{3}.$$

A high cooling rate and a low carbon equivalent favours the formation of white cast iron whereas a low cooling rate or a high carbon equivalent promotes grey cast iron.

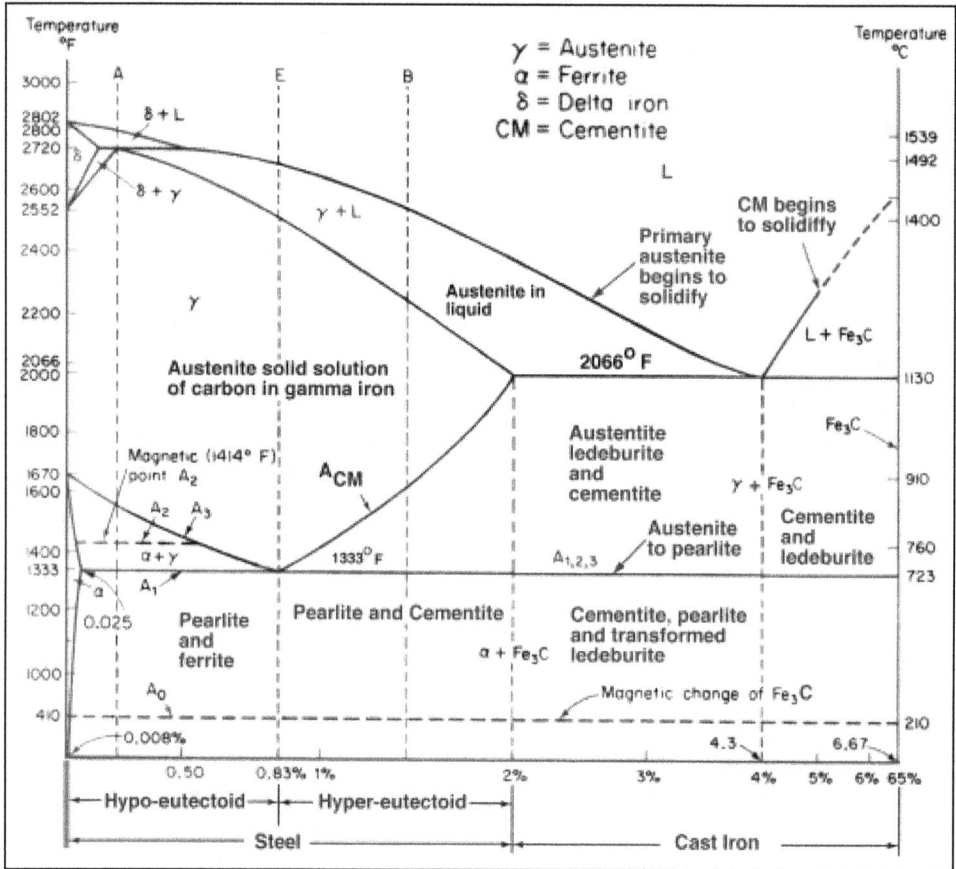

The iron-carbon phase diagram showing the eutectic and eutectoid reactions.

During solidification, the major proportion of the carbon precipitates in the form of graphite or cementite. When solidification is just complete, the precipitated phase is embedded in a matrix of austenite which has an equilibrium carbon concentration of about 2 wt%. On further cooling, the carbon concentration of the austenite decreases as more cementite or graphite precipitates from solid solution. For conventional cast irons, the austenite then decomposes into pearlite at the eutectoid temperature. However, in grey cast irons, if the cooling rate through the eutectoid temperature is sufficiently slow, then a completely ferritic matrix is obtained with the excess carbon being deposited on the already existing graphite. White cast irons are hard and brittle; they cannot easily be machined.

Grey cast irons are softer with a microstructure of graphite in transformed-austenite and cementite matrix. The graphite flakes, which are rosettes in three dimensions, have a low density and hence compensate for the freezing contraction, thus giving good castings free from porosity.

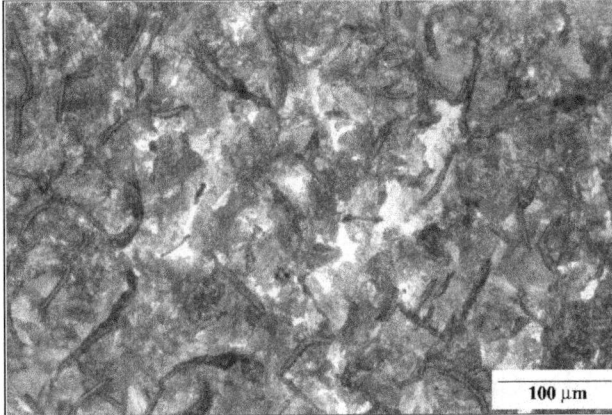

Grey cast iron, Fe-3.2C-2.5Si wt%, containing graphite flakes in a matrix which is pearlitic. The speckled white regions represent a phosphide eutectic.

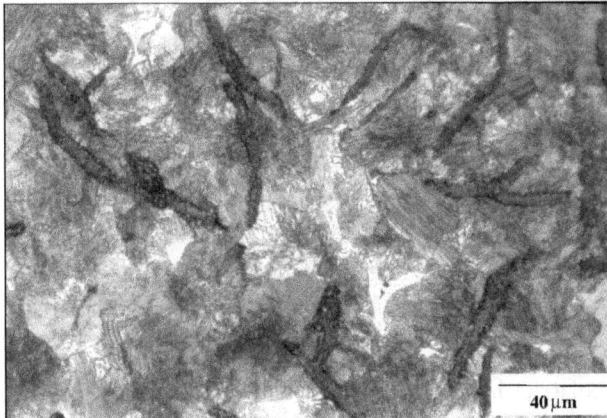

Grey cast iron, Fe-3.2C-2.5Si wt%, containing graphite flakes in a matrix which is pearlitic. The lamellar structure of the pearlite can be resolved, appearing to consist of alternating layers of cementite and ferrite. The speckled white regions represent a phosphide eutectic.

The flakes of graphite have good damping characteristics and good machinability (because the graphite acts as a chip-breaker and lubricates the cutting tools. In applications involving wear, the graphite is beneficial because it helps retain lubricants. However, the flakes of graphite also are stress concentrators, leading to poor toughness. The recommended applied tensile stress is therefore only a quarter of its actual ultimate tensile strength.

Sulphur in cast irons is known to favour the formation of graphite flakes. The graphite can be induced to precipitate in a spheroidal shape by removing the sulphur from the melt using a small quantity of calcium carbide. This is followed by a minute addition of magnesium or cerium, which poisons the preferred growth directions and hence leads to isotropic growth resulting in spheroids of graphite. The calcuim treatment is necessary before the addition of magnesium since the latter also has an affinity for both sulphur and oxygen, whereas its spheroidising ability depends on its presence in solution in the liquid iron. The magnesium is frequently added as an alloy with iron and silicon (Fe-Si-Mg) rather than as pure magnesium.

However, magnesium tends to encourage the precipitation of cementite, so silicon is also added (in the form of ferro-silicon) to ensure the precipitation of carbon as graphite. The ferro-silicon is known as an inoculant. Spheroidal graphite cast iron has excellent toughness and is used widely, for example in crankshafts.

The latest breakthrough in cast irons is where the matrix of spheroidal graphite cast iron is not pearlite, but bainite. This results in a major improvement in toughness and strength. The bainite is obtained by isothermal transformation of the austenite at temperatures below that at which pearlite forms.

Spheroidal Graphite Cast Iron

An illustration of the ductility of spheroidal graphite cast iron. Photograph reproduced from Physical Metallurgy of Engineering Materials, by E. R. Petty, with permission from the Institute of Materials.

The chemical composition of the cast iron is similar to that of the grey cast iron but with 0.05 wt% of magnesium. All samples are etched using 2% nital.

Spheroidal graphite cast iron, Fe-3.2 C-2.5 Si-0.05 Mg wt%, containing graphite nodules in a matrix which is pearlitic. One of the nodules is surrounded by ferrite, simply because the region around the nodule is decarburised as carbon deposits on to the graphite.

Heat Treated Spheroidal Graphite Cast Iron

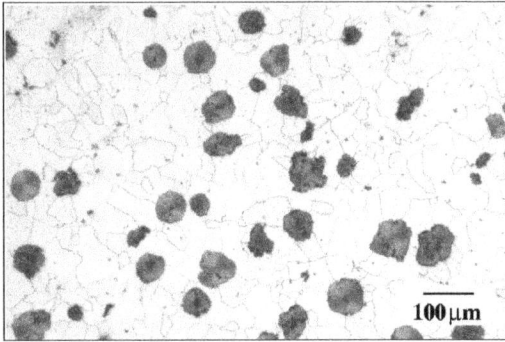

Graphite nodules in a ferritic matrix.

Spheroidal graphite cast iron usually has a pearlitic matrix. However, annealing causes the carbon in the pearlite to precipitate on to the existing graphite or to form further small graphite particles, leaving behind a ferritic matrix. This gives the iron even greater ductility. All samples are etched using 2% nital.

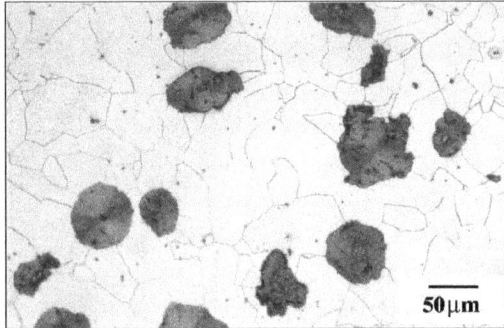

Graphite nodules in a ferritic matrix. Some carbon
deposited during tempering is also visible.

Austempered Ductile Cast Iron

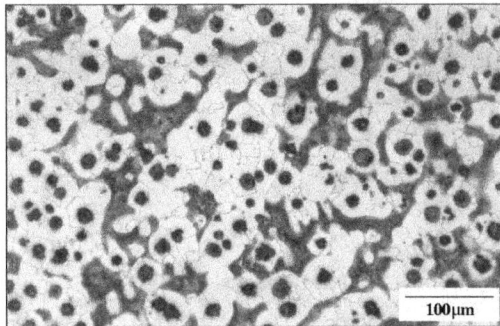

Ductile iron as-cast. Nodules of graphite, pearlite (dark islands) and ferrite (light background).

The chemical composition of the cast iron is Fe-3.52C-2.51Si-0.49Mn-0.15Mo-0.31Cu wt%. All samples are etched using 2% nital. Colour micrographs are produced by first etching with 2% nital, followed by open air heat treatment of the metallographic sample at 270 °C for 3 h. This oxidises the sample and produces interference colours which are phase dependent.

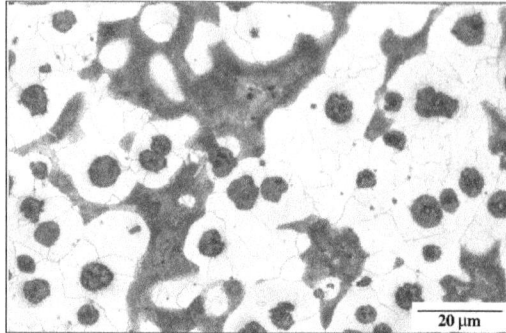

Ductile iron as-cast. Nodules of graphite, pearlite (dark islands) and ferrite (light background).

In order to avoid distortion, the crankshaft for the TVR sportscar is rough-machined after casting, heat-treated to produce the bainitic microstructure, and then properly machined. It is reported to have excellent fatigue properties; its damping characteristics due to graphite reduce engine noise.

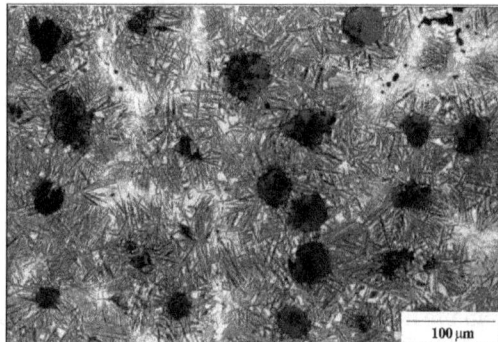

Austenitised 950 °C, austempered 350 °C for 64 min.

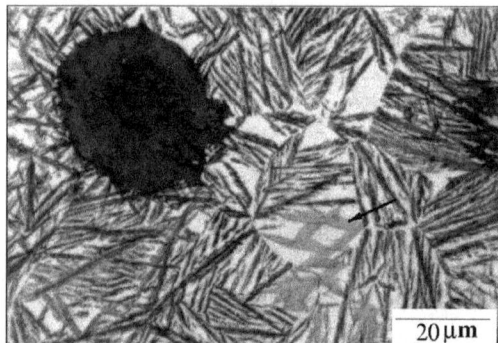

Austenitised at 950 °C, austempered at 350 °C for 64 min.

The Ford Mustang suspension arm was made from austempered ductile iron in order to reduce weight, noise and cost. It was designed using finite element modelling to optimise strength and stiffness. Auminium alloys were considered but rejected because the component would then occupy a much larger space because of their lower strength.

Austempered ductile iron suspension arm for a Ford Mustang Cobra.

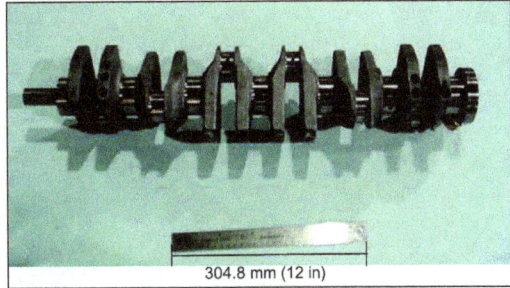

304.8 mm (12 in)

The austempered ductile iron crankshaft for the TVR sportscar.

The truck trailer suspension arm was originally made from welded steel, for use on transportation across the rugged Australian Outback. These failed at the welds and were associated with distortions which led to accelerated deterioration of the tyres. The suspension made from the cast austempered ductile iron has proved to be much more robust.

Blackheart Cast Iron

Blackheart cast iron is produced by heating white cast iron at 900-950 °C for many days before cooling slowly. This results in a microstructure containing irregular though equiaxed nodules of graphite in a ferritic matrix. The term "blackheart" comes from the fact that the fracture surface has a grey or black appearance due to the presence of graphite at the surface. The purpose of the heat treatment is to increase the ductility of the cast iron. However, this process is now outdated since spheroidal graphite can be produced directly on casting by inoculating with magnesium or cerium. All samples are etched using 2% nital.

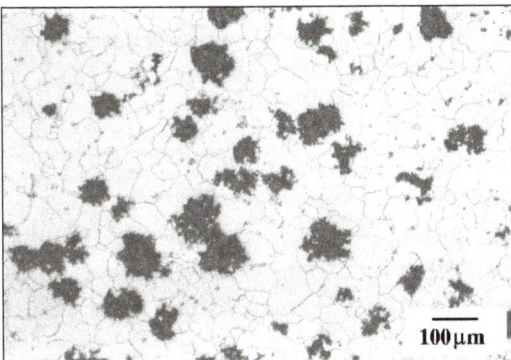

100 μm

50 μm

Blackheart cast iron.

Wear-resistant High-chromium Cast Iron

This cast iron is used in circumstances where a very high wear resistance is desirable. For example, during the violent crushing of rocks and minerals. It contains a combination of very strong carbide-forming alloying elements. Its chemical composition is, therefore, Fe-2.6C-17Cr-2Mo-2Ni wt%.

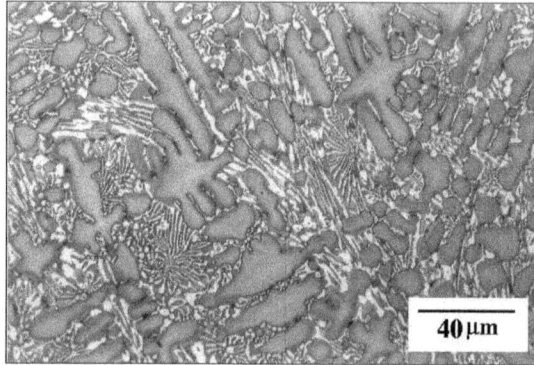

The white phase is a chromium-rich carbide known as M_7C_3. The matrix consists of dendrites of austenite, some of which may have transformed into martensite. There may also be relatively small quantities of other alloy carbides.

All samples are etched using Villela's reagent, which is a mixture of picric acid, hydrochloric acid and ethanol.

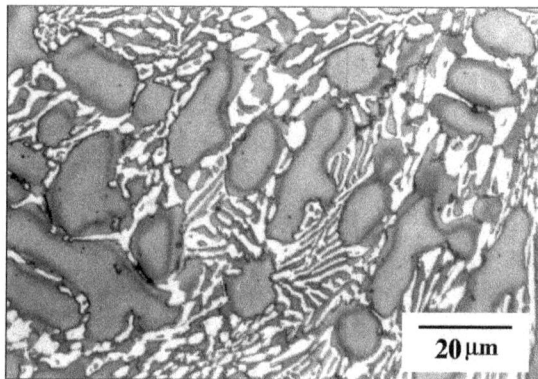

The white phase is a chromium-rich carbide known as M7C3. The matrix consists of dendrites of austenite, some of which may have transformed into martensite. There may also be relatively small quantities of other alloy carbides.

Welding of Cast Irons

The casting process is never perfect, especially when dealing with large components. Instead of scrapping defective castings, they can often be repaired by welding. Naturally, the very high carbon concentration of typical cast irons causes difficulties by introducting brittle martensite in the heat-affected zone of the weld. It is therefore necessary to preheat to a temperature of about 450 °C followed by slow cooling after welding, in order to avoid cracking.

The materials used as fillers during welding usually contain large nickel concentrations so that the resulting austenitic weld metal is not sensitive to the pick-up of carbon from the cast iron. The deposits are soft and can be machined to provide the necessary shape and finish. Of course, nickel is expensive so when making large repairs, the weld gap is first covered ('buttered') with the nickel-rich filler and then the remaining gap is filled with less expensive mild-steel filler metals.

Ironbridge

Ironbridge, made of cast iron.

Ironbridge, made of cast iron.

The world's first bridge made of iron in 1779. The entire structure is made of cast iron.

Cast iron has a "solid feel" and has an appealing appearance. There are many conventional applications of cast iron.

Cast Iron in a Computer Mouse

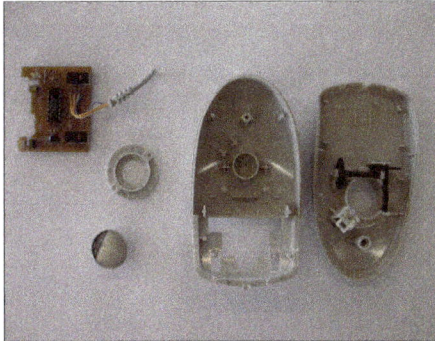

Disection of a computer mouse. The item of interest is the roller ball.

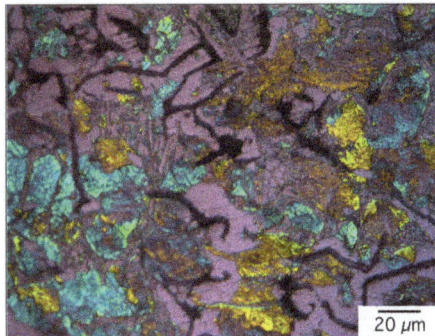

20 μm

The ball is made of cast iron presumably because it is relatively hard.

The microstructure of the roller ball, which is made of cast iron, The flakes of graphite are
surrounded by ferrite, the brown is the peralite, and there is also the product of
the lediburite eutectic which is not clear at this magnification.

Aluminium-silicon Casting Alloys

Aluminium alloys are grouped according to the major alloying elements they contain.
The 4XXX group is alloyed with silicon for ease of casting.Silicon is good in metallic
alloys used for casting. This is because it increases the fluidity of the melt, reduces the
melting temperature, decreases the contraction associated with solidification and is
very cheap as a raw material.

Silicon also has a low density (2.34 g cm^{-3}), which may be an advantage in reducing the
overall weight of the cast component. Silicon has a very low solubility in aluminium; it
therefore precipitates as virtually pure silicon, which is hard and hence improves the
abrasion resistance.

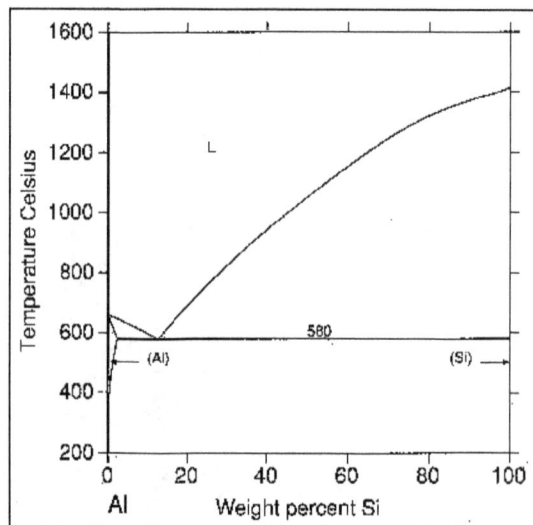

Aluminium-silicon alloys form a eutectic at 11.7 wt% silicon, the eutectic temperature being 577 °C. This represents a typical composition for a casting alloy because it has the lowest possible melting temperature. Al-12Si wt% alloys are therefore common.

Al-12Si (Low Magnification and Unetched)

The dark, semi-circular feature is a casting defect (a pore) caused by the shrinkage of liquid during solidification. The microstructure otherwise consists of grey plates of silicon in a white matrix which is rich in aluminium. Although the alloy is slightly hypoeutectoid in composition, there is evidence that solidification started with primary aluminium dendrites (sections of aluminium dendrite arms are visible). This is because the sample did not solidify under equilibrium conditions. Equilibrium solidification would require painfully slow cooling rates, not achievable in industrial practice.

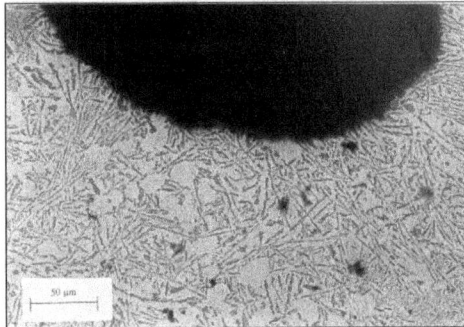

Al-12Si (High Magnification and Unetched)

Shows the coarse silicon plates in an aluminium matrix. The dark feature is a shrinkage pore, a casting defect. Silicon has a diamond crystal structure and is consequently very brittle. Large plates of silicon are, therefore, detrimental to the mechanical properties.

Silicon has a diamond crystal structure and is consequently very brittle. Large plates of silicon are, therefore, detrimental to the mechanical properties. Silicon nucleates on aluminium phosphide particles present in the melt as impurities. The addition of a small amount of sodium to the melt getters the phosphorus, making the nulceation of silicon more difficult. Solidification is therefore suppressed to lower temperatures where the nucleation rate is large. This leads to a remarkable refinement of microstructure.

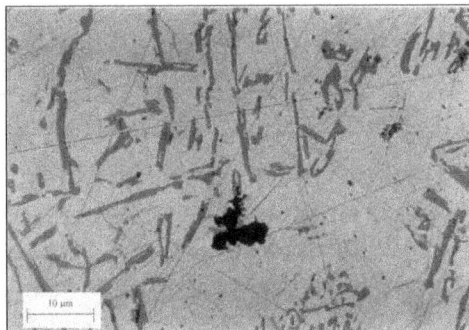

Al-12Si-0.02Na (Low Magnification and Unetched)

The dark feature is a casting defect (a pore) caused by the shrinkage of liquid during so-lidification. The microstructure of this sodium modified alloy is much finer than that of the Al-12Si sample. The silicon particles are hardly visible at this magnification. Notice again the primary dendrites of aluminium, attributed to non-equilibrium solidification.

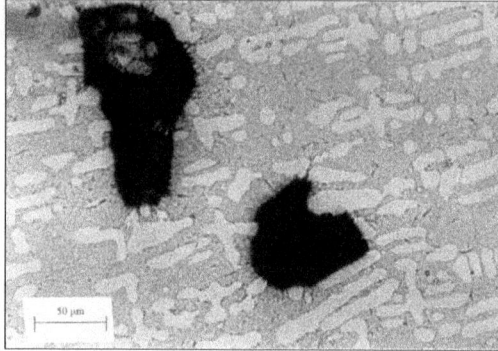

Al-12Si-0.02Na (High Magnification and Unetched)

Greatly refined particles of silicon as the microstructure is modified with sodium.

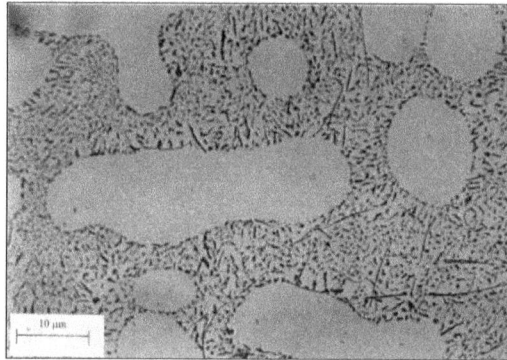

Alloys for Automobile Castings

A typical chemical composition (wt%) for an alloy used in the manufacture of an en-gine-block is as follows:

Si	Cu	Mg	Fe	Mn	Ti	Sr	Zr
7-8	3-4	0.25-0.35	0.0-0.4	0.5	0.00-0.25	trace	0.25

The copper is used for precipitation hardening (Al_2Cu, $Al_5Mg_8Cu_{26}$), should that be nec-essary. Iron is to be avoided if possible, since it can form plate-like precipitates (Al_5Fe-Si) which embrittle the casting and can block the flow of liquid metal in the mould. The strontium, when added deliberately, helps to modify the shape of the silicon, rather as does sodium.

Strontium is preferred to silicon, because the effects of sodium fade relatively rapidly when the liquid metal is held at temperature for a prolonged period before solidification. On the other hand, strontium, by a variety of mechanisms, introduces a greater degree of porosity in the final casting. The following images are of cast aluminium automobile components.

Car steering-knuckle.

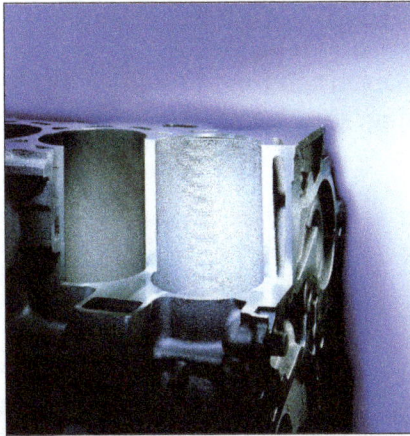

Cast aluminium engine-block made by Hydro Aluminium.

V6 engine-block made from cast aluminium and designed by Hydro Aluminium.

Alloys Ideal for Electronics

A new range of Si/Al alloys containing up to 70 wt% silicon has been developed using spray forming technology. The alloys are 15% lighter than pure aluminium; they have a low, controlled thermal expansion coefficient, a high stifness and high thermal conductivity. They are non-toxic, readily machinable, easily plated with nickel, gold, silver or copper.

Al-Si Coatings: Hot-Press Forming Steels

Very strong steel (1 GPa) can be shaped by hot-press forming. In this, the steel in sheet form is heated to temperatures in the range 900-950 °C for 3-10 min in order to induce it into the austenitic state. It is then formed using a press with water-cooled dies, which simultaneously shape and quench the component to martensite.

The formation of oxide scale during this process is mitigated by applying protective coatings. The micrographs below illustrate what happens to an Al-Si alloy coating containing 7-11 wt% silicon (approximately eutectic composition), as a consequence of the heating associated with hot-press forming. Interdiffusion and reaction between iron, silicon and aluminium leads to the formation Kirkendall voids.

Field emission scanning electron micrographs showing the formation and growth of Kirkendall voids at the coating/steel interface following heat treatment as indicated in the figure captions.

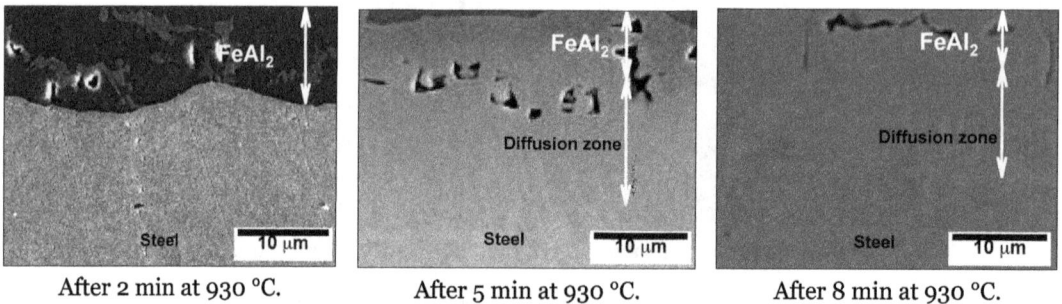

After 2 min at 930 °C.   After 5 min at 930 °C.   After 8 min at 930 °C.

Zinc Casting Alloys

Zinc casting alloys provide a better combination of strength, toughness, rigidity, bearing, performance and economical castability than any other alloy possible. In fact their properties often exceed the ones of other alloys such as aluminium, magnesium, bronze, plastics and other cast irons. For its properties of strength and duration zinc is the perfect choice for saving time and money.

Comparison between Zinc Casting Alloys and Alternative Materials

Designers need to compare materials and examine it in depth at the moment of the choice of the material for the die casting process.

Aluminium

Zinc alloy is more precise than aluminium. Using zinc the designer can create smaller draft angles, smaller and longer cored holes, thinner wall sections are possible. Another important point is that the designer can have a much longer tool life; furthermore zinc has a better machinability and formability but the most important element is that with zinc designers can have lower casting costs.

In terms of avoiding defects zinc alloys is that devices made with this alloy are less likely to leak than those made with aluminium; in fact aluminium tends to expose to porosity and create leaks.

Magnesium

Magnesium is notable for its low density and its price is similar to the aluminium one. When magnesium it is compared to zinc alloys its performances are not so good, in fact in terms of strength to cost ratio and rigidity to cost ratio zinc properties are much more superior than magnesium ones.

Using zinc the designer can save in terms of process costs, can reach a better precision, can have a better corrosion resistance; in addiction zinc has a superior tensile strength and elongation, can create lower draft angles and reach a superior formability.

Machined Steel

Steel is cheaper than zinc alloy but, using zinc, the designer can reduce the process costs reaching a better precision. Steel has a limited design and if the designer need to reproduce complex features he needs to assembly pressings.

We can say that zinc has many advantages but the most important one is that it allows the designer to have a better product saving in terms of cost and time.

Areas where Zinc is Mostly used

Zinc is a material that suits perfectly for many sectors such as:

- Home appliances,
- Automotive,
- Mechanical sector,
- Electronic sector.

We can say that zinc is suitable for different sectors because has many properties that allow the perfect outcome for products in a cost and time saving approach.

Core Benefits of using Zinc

One of the most important benefit of zinc is its accuracy, in fact zinc alloys allow closer tolerances than any other metal or molded plastic. This is one of the major benefits of zinc die casting.

Secondly its machinability because zinc characteristics that are trouble-free decrease machining costs, this is a very competitive issue over other materials. Thin wall capability results smaller, lighter and low cost compared to other metals.

Zinc alloys can be casted with less draft angle than other materials, in fact its components can be casted with zero draft angles that is an advance during the moving mechanical process. All these steps are cost saving. With zinc is also possible to distort parts to achieve the requested shape in a customer - centric way.

In addiction there are inner characteristics that do not need to be compared.

Assembly Operation Reduced

Zinc has many advantages in fact the assembly operations are reduced, the entire part is casted as a single unit without extra expansive manual operations.

Less Material Required

Zinc allows the design of thin wall sections to reduce cost reducing weight and material, thanks to its fluidity, strength and stiffness.

Machining Operation Reduced

Machining operation can be eliminated or reduced thanks to net shape capability of zinc alloy.

Surface Finishing

With zinc is possible to achieve particular aesthetic characteristics.

Eliminate Bearings and Bushings

Zinc allows an amazing design flexibility and with it is possible to reduce other costs such as fabrication costs eliminating small bushing and wear inserts.

Choosing Low, Medium or High Production

There are many different casting processes related to the quantity or size required by the client.

Faster Production

Zinc die casting processes are much faster than aluminium or magnesium ones.

Extended Tool Life

Despite using tools often charges are reduced.

Environmental Harmony

With zinc casting alloys pollution and greenhouse gases are reduced significantly. All those elements sustain the theory that using zinc instead of other alloys such as aluminum or magnesium is better in terms of cost saving.

Of course it's important not to underrate the help that zinc can give not only to performance or structure but in aesthetic terms. Using zinc is possible to create special components, reaching a high aesthetic level given by its finishing.

Finishing

An important advantage of zinc is that it can be finished by a wide variety of techniques because the material is acceptable for many uses in the uncoated state, avoiding in this way the costs for the finishing. If zinc is for an outdoor purpose its structure is disposed to darken with age because it's used in a corrosive external environment.

To avoid this problem it is possible to use some decorative finishes such as electroplated, electropainted, powder coated and wet painted. All this additional elements can be used as a corrosion protection.

Magnesium Alloys

Cast magnesium alloys are manufactured by Die casting, Permanent mold casting and Sand casting methods. Magnesium castings have been produced by a range of casting processes, in a number of foundries around the world for many years. Processes used have included sand casting, low and high pressure die casting, and of recent times using thixocasting. A large range of automotive, consumer goods including computer frames and video camera housings, sporting goods and general castings have been made. Major automotive, aeronautical and transport industry castings are an important current and future growth area for magnesium alloy castings.

Structural Casting Applications

Cast magnesium alloys are being specified at an increasing rate for structural applications, particularly in the area of automotive engineering. The cause of this dramatic growth is a drive by automotive companies to reduce component mass, and to take advantage of specific physical and inherent mechanical properties of magnesium alloys. It has been realised by a number of original equipment manufacturers and suppliers to this industry that superior product performance can be obtained by the use of cast magnesium alloy. A total engineering approach including appropriate material grade selection, component design, casting manufacture, machining and assembly has been adopted by this group. Some manufacturers and first tier suppliers have adopted a magnesium strategy as part of their business operations, where substitution of metal components with magnesium castings is a key part of their future platform planning, involving purchasing and advanced engineering groups. Magnesium casting substitution offers the maximum benefit in the replacement of multi-component fabricated assemblies by a fully integrated die casting. This has been achieved for a number of magnesium seat designs around the world.

AZ91D Magnesium Alloy

When die casting magnesium, design measures and optimal control of processing conditions permit a large number of present applications to be produced from AZ91D grade magnesium alloy. This grade exhibits excellent die castability, with good strength and moderate ductility. It has been considered for decades as the grade of first choice if not discounted by specific property requirements. Typical applications include manual transmission housings, pedal brackets, intake manifolds, steering column and lock housings and mirror brackets to name a few.

AM, AS and AE Series Magnesium Alloys

The AM series of alloys have been developed recently to enhance fracture toughness and hence generate energy absorbing properties. There has been a significant growth in applications for this series as design engineers further exploit the structural casting area.

Typical applications include: steering wheels, seat frames/pans/backs, instrument panels, brackets and firewall beams and door frames/intrusion beams. The AM series are characterised by slightly reduced strengths with higher ductility and impact strength compared to AZ91D grade. For applications involving tong term exposure of stressed components at temperatures exceeding 120 °C, the creep properties need careful consideration and evaluation. As a first measure, design modifications to lower the level of applied stress should be considered. More recent developments, the AS and AE series of alloys, with enhanced creep performance, are available and they should be considered for such applications. These series are based on the addition of either silicon or rare earth elements respectively to promote the formation of finely dispersed particles at the grain boundaries.

Die Casting Production Considerations

Die Life

A significant improvement in die life compared to die casting of aluminium can be expected. This is effected with magnesium because of the heat transfer characteristics and the reduced affinity with iron, resulting in negligible soldering and reduced erosion.

Productivity

Because of the lower heat content of magnesium compared with aluminium, the metal solidifies at a faster rate, generating shorter cycle times, typically by 15-25%. Exceptional dimensional stability of the as-cast product is a particular characteristic of cast magnesium alloys. Frequently, annealing or stress relieving treatments are not required with magnesium, contrary to experiences with some cast aluminium components where some growth continues as natural ageing effects occur over extended times at moderate to elevated temperatures. Machinability is excellent, exhibiting the best characteristics of all the structural materials viz reduced machining time, lower power requirements, longer tool life, excellent surface finish frequently with a single cut and minimal tool build-up with lower overall machining costs.

Sensitivities

During the die casting cycle each part of the casting will develop a microstructure governed by the local solidification rate and pattern. Correct design of the casting and its feeding system are essential to ensure a uniform and directional solidification pattern. Where this is not achieved, it is to be expected that a certain fraction of microporosity will form due to volume contraction during solidification. This will inhibit the achievement of the excellent properties attainable in die cast magnesium. Ductility is a significant process-sensitive parameter with the control of inhomogeneities, defects and process of paramount importance in realising the potential for structural applications.

Classification of Magnesium Alloys

There are three main groups of magnesium alloys (both wrought and cast):

- Magnesium-Manganese Alloys: These alloys have good weldability and are used for manufacturing thin plates.

- Magnesium-Aluminum-Zinc Alloys: These alloys are manufactured by Die casting, Sand casting, Permanent mold casting, Forging and Extrusion. The alloys are heat-treatable.

- Magnesium-Zinc-Zirconium-Thorium Alloys: These alloys have high impact toughness, good corrosion resistance and machinability. The alloys are heat-treatable.

Classification of magnesium alloys was developed by the American Society for Testing and Materials (ASTM).

The designation system uses the following combination of letters and numbers for identification of the alloys. The first two letters indicate the major alloying elements in the alloy according to the following codes:

- A – Aluminum (Al),

- B – Bismuth (Bi),

- C – Copper (Cu),

- D – Cadmium (Cd),

- E – Rare earth elements,

- F – Iron (Fe),

- H – Thorium (Th),

- K – Zirconium (Zr),

- L – Beryllium (Be),

- M – Manganese (Mn),

- N – Nickel (Ni),

- P – Lead (Bb),

- Q – Arsenic (As),

- R – Chromium (Cr),

- S – Silicon (Si),

- T – Tin (Sn),

- Z – Zinc (Zn).

The two letters are followed by two numbers, indicating the concentration of the major alloying elements. The fifth symbol is a letter, signifying the alloy modification.

The alloy code is followed by a designation of temper. The temper designation system of magnesium alloys is similar to the Temper designation of aluminum alloys:

- F –As fabricated,

- O – Annealed,

- H – Cold worked,

- T4 – Solution treatment,

- T5 – Artificial aging,

- T6 – Solution treatment followed by artificial aging.

Example:

Alloy designated as ZE63A-T6 is magnesium alloy, containing 6% (rounded off) of zinc (symbol Z) and about 3% (rounded off) of rare earth elements (symbol E). Modification of the alloy – A. The temper of the alloy is solution treatment followed by artificial aging (T6).

Cast Copper

The use of wrought high conductivity copper for busbars, power cables, domestic wiring and overhead conductors is well established. For non-standard complex shaped components, copper may be cast by any of the traditional methods.

However pure copper is difficult to cast, being prone to surface cracking, shrinkage and internal cavities; it is only specified when castings of very high electrical conductivity are required. The casting characteristics of pure copper can be enhanced by small additions of chromium, beryllium nickel and silicon which enhance the strength without drastically reducing the conductivity.

Cast copper axes and chisels of 99% purity dated 3000 to 3500 BC have been found; clearly the ancients appreciated the hardness and toughness of copper compared to the brittle alternatives of stone and flint.

The increase in copper usage and the beginning of the modern era began in the 19th century with the discovery of electricity, leading to a revolution in communications with telegraph and telephone, followed by fax, leading to internet and satellite systems. This was made possible because of the very high electrical conductivity of copper.

In the mid-1990s, a major improvement in electric motors took place when pressure die-cast copper was used for rotors, resulting in higher efficiency electric motors and showing cast copper at the cutting edge of technology as it has been for 6000 years.

Heat Treatment

CC040A cannot be hardened by heat treatment but may be stress relieved at 250 to 500 °C.

Properties

In the cast condition CC040A has a combination of properties as below:

- Tensile strength: 150 min N/mm^2.

- 0.2% Proof strength: 40 min N/mm^2.

- % Elongation: 25 min.

- Hardness (HV): 40 min.

- Electrical conductivity: 86 -98% IACS.

- Thermal conductivity: 372 W/m °C.

Joining

Soldering and brazing are excellent. Gas shielded arc welding is fair. Oxyacetylene and coated metal arc welding are not recommended.

Machining

The machinability rating is 10% which, with special techniques, is satisfactory. Free-machining brass is 100%.

Resistance to Corrosion

This copper has good resistance to corrosion in industrial and marine atmospheres. It is insensitive to stress corrosion cracking. However, it is susceptible to attack in the presence of ammonia, sulphur, hydrogen sulphide and mercury.

Resistance to Softening

This copper will start to soften at about 150 °C; however it can withstand temperatures up to 250 °C (which may be experienced during short circuits), for a few seconds.

Applications

Components with a complex shape which require the combination of high electrical and thermal conductivity, good corrosion resistance and oxidation resistance include:

- Electrode holders.
- Electrode plates for process industry machinery.
- Rings for process industry machinery.
- Irregular shaped busbars.
- Electrical switchgear.
- Spot welding electrodes.
- Die-cast rotors in high efficiency motors.
- Stressed current-carrying parts for HF welding.
- Contact mechanisms.
- Die-cast battery terminals.
- Terminal lugs.
- High amperage circuit breakers.

Available Casting Processes for CC040A

Sand casting allows the greatest flexibility in casting size and shape and is the most economical casting method if only a few castings are made. Die-casting is more economical above about 50,000 units.

Pressure Die-cast Copper Rotor for Induction Motors

VEM motor with copper rotor cast by Breuckmann.

Electric motors account for two-thirds of all the electricity used in industry, so improving motor efficiency has great economic and environmental benefits. Recent advances in copper die-casting technology have enabled the commercialisation of motors with cast copper rotors in place of aluminium ones. A copper rotor is a rotor made of electrical steel (laminations) where the slots and end rings are filled with copper instead of the traditional material (aluminium). Copper's higher electrical conductivity allows the rotor to conduct electricity more efficiently, resulting in lower resistive losses and lower operating temperature. Motors with copper rotors offer cost savings, weight and space savings, improved thermal capacity, improved steel properties and increased reliability.

Cast Copper Electrode Holder used in Steelmaking

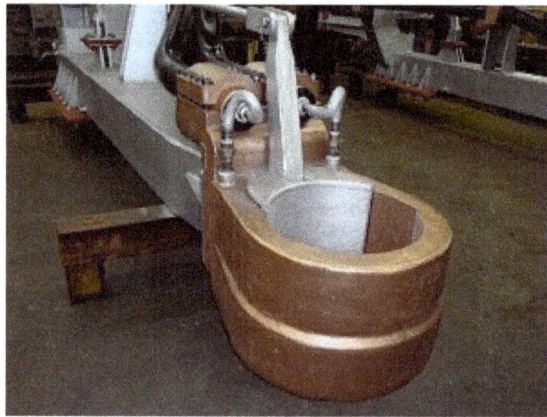

Cast water cooled holder.

The cast water-cooled holder supports a graphite electrode used in electric arc steelmaking. Copper is used because of its high strength, good corrosion resistance, excellent thermal and electrical conductivity which leads to low graphite consumption, high process speed, easy handling, optimal symmetry and maximum efficiency.

Electrical Switchgear

Cast high conductivity components include electrical switchgear such as transformer coils and impedance bonds for railway signalling. The high electrical and thermal conductivity and excellent corrosion resistance of copper is essential for components such as this.

Cast Aluminum Alloys

Aluminum casting alloys offer a range of advantages, particularly a good castability. This includes a relatively high fluidity, low melting point, short casting cycles, relatively low tendency for hot cracking, good as-cast surface finish and chemical stability. In addition, advantages are gained by the specific alloy chosen.

Properties of aluminum and its alloys favorable for casting applications:

- Low melting point.

- Good fluidity of most of the alloys.

- Capability to control grain structure.

- Good surface finish.

- Low solubility of gases (except Hydrogen).

- Ability to be strengthened by heat treatment (precipitation hardening).

Disadvantages of Aluminum Castings:

- High shrinkage (4-8%) and susceptibility to shrinkage defects (shrinkage porosity).

- High hydrogen solubility.

- Susceptibility to hot cracking.

- Low ductility.

The following casting methods are applicable for casting aluminum alloys:

Sand casting, permanent mold casting, die casting, investment casting, centrifugal casting, squeeze casting and continuous casting.

Classification of Cast Aluminum Alloys

Classification of cast aluminum alloys is developed by the Aluminum Association of the United States. Each cast alloy is designated by a four digit number with a decimal point separating the third and the forth digits:

- The first digit indicates the alloy group according to the major alloying element:

  ○ xx.x aluminum, 99.0% minimum,

  ○ xx.x copper (4%...4.6%),

  ○ xx.x silicon (5%...17%) with added copper and/or magnesium,

  ○ xx.x silicon (5%...12%),

  ○ xx.x magnesium (4%...10%),

  ○ xx.x zinc (6.2%...7.5%),

- ○   xx.x tin,

- ○   xx.x others.

- The second two digits identify aluminum alloy or indicate the alloy purity.

In the alloys of the 1xx.x series the second two digits indicate the level of purity of the alloy – they are the same as the two digits to the right of the decimal point in the minimum concentration of aluminum (in percents): 150.0 means minimum 99.50% of aluminum in the alloy, 120.1 means minimum 99.20% of aluminum in the alloy.

In all other groups of aluminum alloys (2xx.x through 9xx.x) the second two digits signify different alloys in the group.

- The last digit indicates the product form: casting (designated by "0") or ingot (designated by "1" or "2" depending on chemical composition limits).

A modification of the original alloy or impurity limits is indicated by a serial letter before the numerical designation. The serial letters are assigned in alphabetical order starting with A but omitting I, O, Q, and X (the letter "X" is reserved for experimental alloys).

Characterization of Aluminum Alloys

Aluminum-copper Cast Alloys (2xx.x series)

- Heat-treatable,

- High strength,

- Low corrosion resistance (susceptible to stress-corrosion cracking),

- Low fluidity,

- Low ductility,

- Susceptible to hot cracks.

Applications

Cylinder heads for automotive and aircraft engines, pistons for diesel engines, exhausting system parts.

Aluminum-silicon-copper/magnesium cast alloys (3xx.x series):

- Heat-treatable,

- High strength,

- Low ductility,

- Good wear resistance,

- Decreased corrosion resistance (in copper containing alloys),
- Good fluidity,
- Good machinability (in copper containing alloys).

Automotive cylinder blocks and head, car wheels, aircraft fittings, casings and other parts of compressors and pumps. Aluminum-silicon cast alloys (4xx.x series):

- Non-heat-treatable,
- Moderate strength,
- Moderate ductility,
- Good wear resistance,
- Very good cast properties,
- Good corrosion resistance.

Pump casings, thin wall castings, cookware. Aluminum- magnesium cast alloys (5xx.x series):

- Non-heat-treatable,
- High corrosion resistance,
- Good machinability,
- Good appearance when anodized,
- Moderate cast properties.

Sand cast parts. Aluminum-zinc cast alloys (7xx.x series):

- Heat-treatable,
- Good dimensional stability,
- Good corrosion resistance,
- Poor cast properties,
- Good machinability (in copper containing alloys).

Aluminum-tin cast alloys (8xx.x series):

- Non-heat-treatable,
- Low strength,
- Very good wear resistance,
- Good machinability.

Monometal (solid) and bi-metal slide bearings for internal combustion engines and other slide bearings applications.

Cast Steel

Cast steel is a ferrous alloy with a maximum carbon content of approximately 0.75%. Steel castings are solid metal objects produced by filling the void within a mold with liquid steel. They are available in many of the same carbon and alloy steels that can be produced as wrought metals. Mechanical properties for cast steel are generally lower than wrought steels, but with the same chemical composition. Cast steel compensates for this disadvantage with its ability to form complex shapes in fewer steps.

Steel castings are solid metal objects produced by filling the void within a mold with liquid steel.

Properties of Cast Steel

Cast steels can be produced with a wide range of properties. The physical properties of cast steel change significantly depending on chemical composition and heat treatment. They are selected to match performance requirements of the intended application.

- Hardness: The ability of a material to withstand abrasion. Carbon content determines the maximum hardness obtainable in steel, or hardenability.

- Strength: The amount of force necessary to deform a material. Higher carbon content and hardness result in steel with higher strength.

- Ductility: The ability of a metal to deform under tensile stress. Lower carbon content and less hardness result in steel with higher ductility.

- Toughness: The ability to withstand stress. Increased ductility is usually associated with better toughness. Toughness can be adjusted with the addition of alloying metals and heat treatment.

- Wear resistance: The resistance of a material to friction and use. Cast steel exhibits similar wear resistance to that of wrought steels of similar composition. The addition of alloying elements such as molybdenum and chromium can increase wear resistance.

- Corrosion resistance: The resistance of a material against oxidization and rust. Cast steel exhibits similar corrosion resistance to that of wrought steel. High-alloy steels with elevated levels of chromium and nickel are highly oxidation resistant.

- Machinability: The ease at which a steel casting can change shape by removing material through machining (cutting, grinding, or drilling). Machinability is influenced by hardness, strength, thermal conductivity, and thermal expansion.

- Weldability: The ability of a steel casting to be welded without defects. Weldability is primarily dependent on the steel casting's chemical composition and heat treatment.

- High-temperature properties: Steels operating at temperatures above ambient are subject to degraded mechanical properties and early failure due to oxidation, hydrogen damage, sulfite scaling, and carbide instability.

- Low-temperature properties: The toughness of cast steel is severely reduced at low temperatures. Alloying and specialized heat treatments can improve a casting's ability to withstand loads and stresses.

Chemical Composition of Cast Steel

The chemical composition of cast steel has a significant bearing on performance properties and is often used to classify steel or assign standard designations. Cast steels can be broken into two broad categories—carbon cast steel and alloyed cast steel.

Carbon Cast Steel

Like wrought steels, carbon cast steels can be classified according to their carbon content. Low carbon cast steel (0.2% carbon) is relatively soft and not readily heat-treatable. Medium carbon cast steel (0.2–0.5% carbon) is somewhat harder and amenable to strengthening by heat treatment. High carbon cast steel (0.5% carbon) is used when maximum hardness and wear resistance are desired.

Alloyed Cast Steel

Alloyed cast steel is categorized as either low- or high-alloy. Low-alloy cast steel ($\leq 8\%$ alloy content) behaves similarly to normal carbon steel, but with higher hardenability. High-alloy cast steel ($> 8\%$ alloy content) is designed to produce a specific property, such as corrosion resistance, heat resistance, or wear resistance.

Common high-alloy steels include stainless steel (> 10.5% chromium) and Hadfield's manganese steel (11–15% manganese). The addition of chromium, which forms a passivation layer of chromium oxide when exposed to oxygen, gives stainless steel excellent corrosion resistance. The manganese content in Hadfield's steel provides high strength and resistance to abrasion upon hard working.

Cast Steel Grades

Steel grades have been created by standards organizations such as ASTM International, the American Iron and Steel Institute, and the Society of Automotive Engineers to classify steels with specific chemical compositions and resulting physical properties. Foundries may develop their own internal grades of steel to meet user demand for specific properties or to standardize specific production grades.

The specifications for wrought steels have often been used to classify different cast alloys by their principal alloying elements. However, cast steels do not necessarily follow wrought steel compositions. The silicon and manganese contents are frequently higher in cast steels compared with their wrought equivalents. In addition to their predominantly higher levels of silicon and manganese, alloyed cast steels use aluminum, titanium, and zirconium for de-oxidation during the casting process. Aluminum is predominantly used as a de-oxidizer for its effectiveness and relative low cost.

Cast Steel Production

The practice of casting steel dates back to the late 1750s, much later than the casting of other metals. The high melting point of steel, and the lack of technology available to melt and process metals, delayed the development of a steel casting industry. These challenges were overcome by advances in furnace technology.

Furnaces are refractory lined vessels that contain the "charge," which is the material to be melted, and provides energy for melting. There are two furnace types used in a modern steel foundry: electric arc and induction.

Electric Arc Furnace

The electric arc furnace melts batches of metal referred to as "heats" by means of an electric arc between graphite electrodes. The charge passes directly between the electrodes, exposing it to thermal energy from the ongoing electrical discharge. Electric arc furnaces follow a tap-to-tap operating cycle:

- Furnace charging: Load of steel scrap and alloys are added to the furnace.

- Melting: Steel is melted by supplying energy to the furnace interior. Electrical energy is supplied through graphite electrodes and is usually the largest contributor in steel melting operations. Chemical energy is supplied through oxy-fuel burners and oxygen lances.

- Refining: Oxygen is injected to remove impurities and other dissolved gasses during the melting process.

- De-slagging: Excess slag, which often contains undesirable impurities, is removed from the bath prior to tap out. De-slagging can also take place within the ladle prior to pouring.

- Tapping (or tap out): Metal is removed from the furnace by tilting the furnace and pouring the metal into a transfer vessel such as a ladle.

- Furnace turn-around: Tap out and preparation are completed for the next furnace charge cycle.

The electric arc furnace melts batches of metal using graphite electrodes; the charge passes directly between the electrodes, exposing it to thermal energy.

Continuous additional steps are often taken at various stages in this process to further de-oxidize the steel and to remove slag from the metal prior to pouring. The steel's chemistry may need to be adjusted to account for alloy depletion during an extended tap-out.

Induction Furnace

An induction furnace is an electrical furnace where heat energy is transferred by induction. A copper coil surrounds the nonconductive charge container, and an alternating current is run through the coil to create an electromagnetic induction within the charge.

Induction furnaces are capable of melting most metals, and they can be operated with minimal melt loss. The disadvantage is that little refining of the metal is possible. Unlike an electric arc furnace, the steel cannot be transformed.

Modern steel foundries frequently utilize recycled steel scrap to reduce the cost and environmental impact of casting production. Obsolete cars, mechanical components, and similar items are separated, sized, and shipped to foundries as scrap. This is combined with internal scrap generated in the casting process and combined with various alloying elements to charge the melting furnace.

An induction furnace is an electrical furnace where heat energy is transferred by induction, and can operate with minimal melt loss.

Heat Treatment

After the casting is solidified, removed from the mold, and cleaned, the physical properties of cast steel are developed by proper heat treatment.

- Annealing: Heating steel castings to a specific temperature, holding for a specific period of time, and then slowly cooling.

- Normalizing: Similar to annealing, but steel castings are cooled in open air, sometimes with fans. This helps the castings to achieve higher strengths.

- Quenching: Similar to normalizing, but cooling takes place at a much more rapid pace using forced air. Water or oils are used as the quench medium.

- Tempering (or stress relieving): Technique used to relieve internal stresses from within castings. These stresses can appear from the casting process, or during strengthening or hardening heat treatments such as normalizing or quenching. Stress relieving involves heating the castings to a temperature well below the annealing temperature, holding it at that temperature, and then slow cooling.

Cast Steel Inspection

Steel castings are often subject to inspections to verify specific physical properties such as dimensional accuracy, cast surface finish condition, and internal soundness. In addition, chemical composition must also be inspected. Chemical composition is dramatically affected by minor alloying elements added to the material. Cast steel alloys are susceptible to variations of their chemical composition, so chemical analyses are required to verify the exact chemical composition prior to casting. A small sample of molten metal is poured into a mold and analyzed.

Dimensional Accuracy

Dimensional inspections are carried out to make sure that the castings produced meet the customer's dimensional requirements and tolerances including allowances for machining. It may sometimes be necessary to destroy sample castings to take measurements of interior dimensions.

Surface Finish Condition

Casting surface finish inspections are employed to explore the aesthetic appearance of castings. They look for flaws in the surface and sub-surface of the castings that may not be obvious visually. The surface finish of a steel casting may be influenced by the type of pattern, molding sand, and mold coating used, as well as the weight of the casting and methods of cleaning.

Surface finish inspection looks for flaws in the surface and sub-surface of steel castings.

Internal Soundness

All castings have some level of defects present, and the soundness specification determines the acceptable defect threshold. Over-specification of the maximum allowable defect level will lead to higher scrap rates and higher casting costs. Under-specification of the maximum allowable defect level can lead to failure.

Three common internal defects that occur in steel castings are:

1. Porosity: Voids in the steel casting that are characterized by smooth, shiny interior walls. Porosity is generally a result of gas evolution or gas entrapment during the casting process.

2. Inclusions: Pieces of foreign material in the casting. An inclusion can be metallic, intermetallic, or non-metallic. Inclusions can come from within the mold (debris, sand, or core materials), or can travel into the mold during the pouring of the casting.

3. Shrinkage: Vacancy or area of low density typically internal to the casting. It is caused by a molten island of material that does not have enough feed metal to supply it during the solidification process. Shrinkage cavities are characterized by a rough crystalline interior surface.

Chemical Analysis

Chemical analysis of cast steels is usually performed by wet chemical analysis methods or spectrochemical methods. Wet chemical analysis is most often used to determine the composition of small specimens, or to verify product analysis post-production. Contrastingly, analysis with a spectrometer is well-suited to the routine and rapid determination of the chemical composition of larger samples in a busy production foundry environment. Foundries can undertake chemical analysis at both the heat and product level.

Heat Analysis

During heat analysis, a small sample of liquid cast steel is ladled from the furnace, allowed to solidify, and then analyzed for chemical composition using spectrochemical analysis. If the composition of alloying elements is not correct, quick adjustments can be made in the furnace or ladle prior to casting. Once correct, a heat analysis is generally considered to be an accurate representation of the composition of the entire heat of metal. However, variations in chemical composition are expected due to segregation of alloying elements, and the time it takes to pour off the heat of steel. Oxidation of certain elements may occur during the pouring process.

During heat analysis, a sample of liquid cast steel is ladled from the furnace, allowed to solidify, and then analyzed for chemical composition using spectrochemical analysis.

Product Analysis

Product analysis is performed for specific chemical analysis verification, as the composition of individual castings poured may not entirely conform to the applicable specification. This can happen even if the product was poured from a heat of steel where the heat analysis was correct. Industry practices and standards do allow for some variation between heat analysis and product analysis.

Cast Steel Testing

A variety of mechanical properties can be achieved for carbon and alloy steel castings by altering the composition and heat treatments of cast steels. Foundries utilize specialized testing methods to check mechanical properties prior to product completion.

When it comes to cast steel testing, there are two types of testing used in industry: destructive and non-destructive testing. Destructive testing requires the destruction of a test casting to visually determine the internal soundness of a part. This method only provides information on the condition of the piece tested, and does not ensure that other pieces will be sound. Non-destructive testing is employed to verify the internal and external soundness of a casting without damaging the casting itself. Once the casting passes the tests, it can be used for its intended application.

Tensile Properties

Tensile properties for steel castings are an indication of a casting's ability to withstand loads under slow loading conditions. Tensile properties are measured using a representative cast sample that is subject to controlled tensile loading—pulling forces exerted on either end of the tensile bar—until failure. Upon failure, tensile properties are examined.

Tensile Properties	
Properties	Description
Tensile Strength	Stress required to break a casting in tension, or under a stretching load.
Yield Strength	Point at which a casting begins to yield or stretch and demonstrate plastic deformation while in tension.
Elongation (%)	Measure of ductility, or the ability of a casting to deform plastically.
Reduction of area (%)	Secondary measure of a casting's ductility.
	Demonstrates the difference between original cross-sectional area of the tensile bar, and area of smallest cross section after failure in tension.

Bend Properties

Bend properties identify a casting's ductility by using a rectangular representative sample bent around a pin to a specific angle. The resulting bent bar is observed to check for objectionable cracking.

Impact Properties

Impact properties are a measure of toughness resulting from testing the energy required to break a standard notched sample. The more energy required to break the sample, the tougher the cast material.

Hardness

Hardness is a measure of a casting's resistance to penetration using indentation tests. It is a property that indicates wear and abrasion resistance of cast steels. Hardness testing can also provide an easy, routine method to test for indications of tensile strength in a production environment. A hardness scale test result will normally correlate closely with tensile strength properties.

Titanium Alloy

Use of titanium and titanium alloys started in the early 1950s. Soon, it becomes very popular with the aerospace, energy, and chemical industries around the globe. Titanium and its alloys are the best material choice for various critical applications because of their high strength-to-weight ratio, excellent mechanical properties, and corrosion resistance features. Titanium alloys are used for many critical hi-tech applications, such as rotating and static gas turbine engine components and parts of aircraft engines.

Other application areas of titanium alloys are:

- Nuclear power plants.

- Food processing plants.

- Oil refinery heat exchangers.

- Marine components.

- Medical protheses.

Almost 80% of all the titanium produced worldwide is used in the aerospace industries.

Buying Tips

Dimensions, performance features, and production processes are three things that should be analyzed properly before selecting titanium and titanium alloys.

- Dimensions: Outer diameter (OD), inner diameter (ID), overall length, and overall thickness are important dimensions.

- Performance features: It includes resistance to corrosion, heat, and wear.

- Production processes: Most materials are cast, wrought, extruded, forged, cold-finished, hot-rolled, or formed by compacting powdered metals or alloys.

Advantages of Titanium Alloys

Titanium is a light, strong, lustrous, corrosion-resistant transition nonferrous metal. It can be easily alloyed with other elements/metals including iron, aluminum, vanadium, molybdenum and others, for producing strong lightweight alloys for aerospace and other demanding applications. Titanium's advantages can be summarized as follows:

- Elevated temperature 350 °F-1000 °F service capability.

- Excellent fatigue and fracture resistance.

- Excellent strength-to-weight ratio.

- Compatibility with carbon/epoxy materials.

- It is used as part of the containers of batteries and as anode in alkaline batteries.

- Biocompatibility.

- Superior oxidation and corrosion resistance.

- Non-magnetic character.

- Fire resistance.

- Short radioactive half life.

Types of Titanium Alloys

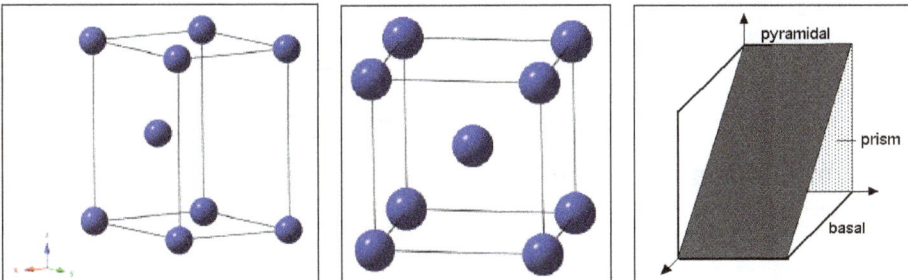

Crystal structure of α-titanium.

Titanium alloys are classified as Alpha (a), Alpha-Beta (a-ß), and Beta (ß) alloys on the basis of alloying elements they contain. The following table explains and compares these three titanium alloys:

Alpha Alloys (α)

Alpha alloys commonly have creep resistance superior to beta alloys. Alpha alloys are suitable for somewhat elevated temperature applications. They are also sometimes used for cryogenic applications. Alpha alloys have adequate strength, toughness, and weldability for various applications, but are not as readily forged as many beta alloys. Alpha alloys cannot be strengthened by heat treatment.

Alpha-Beta Alloys (α-β)

Alpha-Beta alloys have chemical compositions that result in a mixture of alpha and beta phases. The beta phase is normally in the range of 10 to 50% at room temperature. Alloys with beta contents less than 20% are weldable.The most commonly used titanium alloy is Ti-6Al-4V, an alpha + beta alloy. While Ti-6Al-4V is fairly difficult to form other alpha + beta alloys normally have better formability.

Beta Alloys (β)

Beta alloys have good forging capability. Beta alloy sheet is cold formable when in the solution treated condition. Beta alloys are prone to a ductile to brittle transition temperature. Beta alloys can be strengthened by heat treatment. Typically beta alloys are solutioned followed by aging to form finely dispersed particles in a beta phase matrix. Followings are the few common titanium alloys according to the above classifications:

Alpha Alloys (α)	Alpha-Beta Alloys (α-β)	Beta Alloys (β)
Ti-2.5Cu	Ti-6Al-4V	Ti-13V-11Cr-3Al
Ti-5Al-2.5Sn	Ti-6Al-6V-2Sn	Ti-8Mo-8V-2Fe-3Al
Ti-8Al-1V-1Mo	Ti-6Al-2Sn-2Zr-2Cr-2Mo	Ti-10V-2Fe-3Al
Ti-6242	Ti-3Al-2.5V	Ti-15-3
Ti-6Al-2Nb-1Ta-0.8 Mo	Ti-8Al-1Mo-1V	-----
Ti-5Al-5Sn-2Zr-2Mo	-----	-----

Applications of Titanium Alloys

- Artillery (howitzers).

- Turbines (power generation).

- Manned and unmanned aircraft (commercial and military aircraft, rotorcraft).

- Automotive (motorcycles, performance automobiles).

- Military vehicles (tanks, hovercraft).

- Naval and marine applications (surface vessels, subs).

- Sports equipment (bicycle frames, golf clubs).

- Chemical processing plants (petrochemical, oil platforms).

- Consumer electronics (batteries, watches).

- Pulp and paper industry (washing and bleaching systems).

- Medical devices (implants, instruments).

- Architecture (sculptures).

Investment Casting of Titanium Alloys

Investment casting is the proven casting method of choice to ensure the highest possible quality of end product by using a shell system to help control the surface finish and detailing of the casting.

There are a wide range of titanium cast alloys to choose from however depending the application it important to consider the alloy choice carefully.

The main difficulty with casting titanium alloys is their reactivity with common elements in air like oxygen and nitrogen. The investment casting process is the only metal casting process that can produce complicated shapes in high temperature alloys to a very high quality standard.

The shell system and the shell technology, including the control, maintenance, composition and operation, have the biggest effect on the quality of the casting. The shell system comprises a face coat system and a back-up system to produce the shell mould. Apart from the quality of the wax pattern, the composition of the slurry and the stucco's used for the shell manufacture as well as the physical properties of the slurries, controls the surface finish of the casting and the definition of the features of the casting.

Titanium castings are successfully being implemented as cost-effective alternatives to forged and wrought products for high performance and increasingly cost-sensitive applications such as military and commercial air craft airframe structures. In some instances, these castings have been produced for half the cost of comparable forged and machined parts. For most of the last two decades, investment casting has been the preffered processing route for sophisticated titanium castings.

Table: Examples of cast titanium alloys.

Ally Designation	Preferred property	Application
CP-Ti (DIN 17865)	Corrosion resistance	Chemical industry
Ti-64 (EN 3352)	Strength, weight	Universal application
Ti-6242 (WI.3.7141)	Temperature capability, creep	Moderately elevated temperatures
TIMETAL 834	Temperature capability, creep	High temperatures
γ -TiAl	Weight temperature, creep	Very high temperatures

Table: Typical mechanical properties of cast titanium alloys.

Alloy Designation	UTS [MPa]	YS [MPa]	EL.[%]	Max. T [°C]
CP-Ti (DIN 17865)	350	280	15	350
Ti-64 (EN 3352)	880	815	5	350
Ti-6242 (WL 3.7141)	860	760	6	450
TIMETAL 834	1020	900	4	600
γ -TiAl	500-600	400-500	1-2	800

Various investment cast parts for applications in the low
temperature section of a gas turbine engine.

The most widely used titanium alloy nowadays is Ti-6Al-4V. This alloy possesses an excellent combination of strength, toughness and good corrosion resistance and finds application in aerospace, pressure vessels, aircraft compressor blades and discs, surgical implants etc. Aluminum stabilizes the hexagonal close-packed (hcp) α phase, and vanadium, being body-centered cubic (bcc), stabilizes the β phase. Because of high melting point and excessive reactivity of the melt with crucibles, melting and pouring of titanium alloys have to be performed under vacuum. Due to the high cost of titanium, the use of net-shape or near-net-shape technologies receive an increasing interest considering the large cost saving potential of this technology in manufacturing parts of complex shapes. Precision (investment) casting is by far the most fully developed net-shape technology compared to powder metallurgy, superplastic forming and precision forging. Production of precision castings of titanium alloys was considerably increased during last years due to significant cost savings compared with complicated process of machining. When Ti-6Al-4V is slowly cooled from the β region, α begins to form below the β transus temperature that is about 980 °C. The kinetics of β→α transformation upon cooling strongly influences properties of this alloy. Contrary to wrought material, however, the possibilities to optimize the properties via the microstructural control are limited for cast parts to purely heat treatments. For many alloys mechanical properties of castings are inherently lower than those of wrought alloys. Nevertheless, heat treatment of titanium castings yields mechanical properties comparable, and often superior, to those of wrought products.

References

- Casting-alloys: themetalcasting.com, Retrieved 3 May, 2019

- Alloy-die-casting, custom-manufacturing-fabricating: thomasnet.com, Retrieved 13 February, 2019

- Cast.iron, adi: msm.cam.ac.uk, Retrieved 17 March, 2019

- The-advantages-of-zinc-casting-alloys: bruschitech.com, Retrieved 23 June, 2019

- Cast-copper, conductivity-materials, about-copper: copperalliance.org.uk, Retrieved 12 April, 2019

- Cast-aluminum-alloys, properties-of-aluminum-and-its-alloys-favorable-for-casting-applications: substech.com, Retrieved 21 March , 2019

- Cast-steel: reliance-foundry.com, Retrieved 13 June, 2019

- Titanium-alloy: themetalcasting.com, Retrieved 24 March , 2019

Post-casting Processing

There are diverse processes which take place after casting has been completed. A few of these processes are heat treatment, machining of the cast materials, and painting and finishing. These post casting processes have been thoroughly discussed in this chapter.

Casting finishing is the process of taking cooled, molded cast assemblies and preparing them for use. Depending on the process used, (nobake, green sand, investment, etc.) the castings will require various levels of finishing. The type of metal cast also plays a part in the finishing process.

In general, the castings must first be removed from the tree assembly they were cast with. Cut-off saws are usually used for this task. Castings sometime require additional areas of metal to be added to balance the metal flow during pouring. These areas are called risers. Risers must be removed by either cutting them off or knocking them off by force, such as using a sledge hammer or Impactor.

Once the castings are separated from the trees or risers, all flash and parting line metal must be removed. Each of these is a result of casting through the use of compacted molds. Depending on the casting finish quality, finishing may involve several steps of grinding from coarse to fine.

After finishing, castings are inspected for surface quality. Inspection can be performed manually by visual checking, manually by template comparison or by an automated inspection station.

Post-casting Finishing Processes used in Die Cast Components

All the die cast parts undergo a series of processes. While die casting, the metal molds or dies are preheated and coated in a die release agent before the molten metal is injected. However, the casted components need to undergo numerous coatings, finishes, painting, and polishing before shipping. This process ensures that the finished parts meet the necessary cosmetic requirements, protect against corrosion, and improve its wear resistance. Carrying out this finishes is always in accordance with the die design.

The preparation of the surface requires considering the functional design features of the part. The areas of the part such as the edges may often need special polishing, shave trim, chromate coating, and painting. The geometry of design of specific part features largely influences the type and quality of the final finish. By carrying out small-scale modifications of crucial surfaces and edges, the costs can be reduced along with lesser surface preparation before the final coat is applied.

Aluminum, zinc, and magnesium die castings are subjected to about three of the post casting finishing steps. This is carried out depending on the durability, cosmetic appearance, and protection of the part. Some of the post finishing steps carried out are mentioned below:

- Post-Trim Deburring – Several vibratory processes are done to round sharp edges, smoothen surfaces, eliminate burrs, and to release flash and debris. Mechanical deburring is done in most of the die castings before they are subjected to post-trim finishing.

- Surface Conversion Coating – Conversion coating is an important post-casting finishing step that will fully remove any oil, die cast part release agents, and other contaminants remaining on the surface. Conversion coating is carried out as a preparatory process, as it can act as a primer when the part is intended to undergo final painting.

- Combined Conversion Coatings – Combined conversion coatings are applied when a die cast component requires a specific function such as corrosion resistance, durability, and enhancements in its cosmetic appearance. It can be used as a replacement for surface conversion coating, as it can also serve as a paint base or the final finish.

- Final Cosmetic Surface Finish – A final painting or plating of the die cast components is carried out for cosmetic purposes. Maintaining this appearance, a final applied finish will help in increasing the corrosion resistance, as well as boost heat dissipation for improved insulation properties and surface performance.

- Powder coating, liquid paint polyurethane, and water-based finishes are some of the final finished applied. An important part of cosmetic surface finish is component masking that ensures that the areas that do not receive finish coating remain protected.

- Painting – Painting is done by either powder or liquid. Non-solvent based powder coatings have the advantage of being non-toxic, which enable safe waste disposal, thus being environmentally friendly. It provides uniformity in the surface finish, provides durability, and is available in several surface textures such as from matte to semi-gloss.

The most common final color finishes used for die casting components are polyurethane and other wet paint chemistries. Compared to powder coating, the production costs and lead times are much lower when liquid paints are used in the process.

Casting Cleaning Technique

The solidified part is also known as a casting, which is ejected or broken out of the mold to complete the process. After the molten metal has been poured into the mould, it is permitted to cool and solidify. When the casting is solidified, it is removed from the sand in the moulding box. This operation is called Shake out. Mechanical shakes out can be used in large-scale works. Casting, when taken out of the mould, is not in the same condition in which they are desired since they have sprues, risers, gates, etc. attached to them. Besides, they are not completely free from sand particles. The operation of cutting off the unwanted parts, cleaning and finishing the casting is known as fettling. Defects such as blow holes, gas holes, cracks, warping, deformation may often occur in castings. Such defective castings cannot be rejected outright for the reasons of economy and they are therefore repaired by suitable means which include various types of welding, soldering, resin impregnation, epoxy filling, metal spraying etc. Deformed or wrapped castings can be straightened in a press by applying pressure. These entire defects can visible only after casting cleaning process, hence casting cleaning process is very important step in casting techniques economics.

Water Jet Cleaning

It is generally used for cleaning the critical areas of casting. This process generally used after the shot blasting process, shot peening or any machining process to remove core sand particles and burr present inner side of casting.

The nozzle diameter is selected as per the required area cleaning and then high pressurized water and cleaning solvent is forced by pump into required (cleaning) area. It is very useful for complicated castings. It is very accurate cleaning method and required degree of cleanliness. By using water jet cleaning we will get desired degree of cleanliness value called as Millipore value. Millipore is Italian terminology defined as maximum value of the impurities present in the component after cleaning the casting.

Water jet cleaning uses a stream of supersonic water at pressures between 30,000 to 60,000 PSI (207–414 MPa) to quickly remove difficult coatings and debris from a material substrate. Water jet cleaning is different than conventional cleaning processes in that it uses much higher water pressures which require unique pump, hydraulic and control systems. Very hard coatings are essentially eroded from the substrate by the high pressure water droplets, while brittle coatings are fractured and spalled. Water jet cleaning systems are used to remove many coatings, including grease, adhesive sand, epoxies, rubber, felt metal, resin composites, paints, thermal spray coatings (including ceramics, metallic's, abradables, and cermets).

Attributes of Water Jet

- High performance, high production.

- Lightweight, simple to operate and maintain.

- Combines the power of a solid pencil jet with the large area coverage of a fan jet.

- Rugged and dependable.

- Rotor and insert assemblies are specific to your exact flow and pressure requirement.

Applications

- Casting Cleaning.

- Scale removal.

- Surface preparation.

- Root cutting.

Multinozzle Cleaning

The multi nozzle system with Jet cleaning can, in principle, be used wherever there is a need for economical cleaning of tanks and storage chambers in sewage plant. The Multinozzle System is particularly suited for automatic cleaning of tank walls as shown in figure. Casting surface, floors used for storm water retention, over flow and collection reservoirs as well as for the cleaning of storage chambers within sewage treatment systems. In addition, air input necessary for the cleaning process regenerates the sewage water and alleviates any unpleasant odor problems. The multi jet circulates the storm water from the bottom of the tank and passes it to an ejector nozzle as shown in figure.

Tank Cleaning.

A multi-nozzle electro-spray system was developed for application to gas cleaning from fine particles. Such a method, known as electrostatic scrubbing, is usually employed for removal of fine particles (smaller than a few micrometers in diameter) from flue gases. Such particles are very difficult to remove by conventional systems such as inertial scrubbers, bag filters or electrostatic precipitators as shown in figure. The operation of electrostatic scrubbers is based on Coulomb attraction of charged particles by oppositely charged droplets. Electrostatic scrubbers have been proposed for many years to solve the problem of small particles removal, but the main issue of high water consumption by these systems is still unsolved.

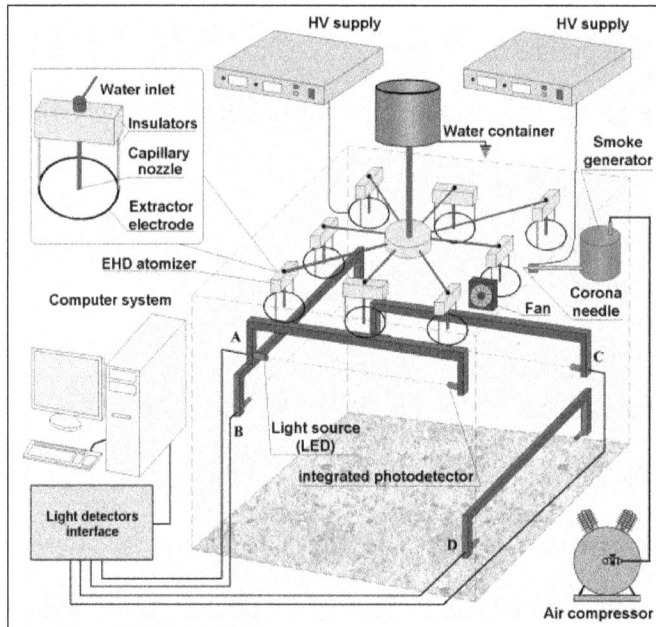

Multi-nozzle electro-spray system.

Surface Cleaning

The removal of organic material such as lubricating oils or other processing aids, for example release agents, is routinely carried out as part of most multi-stage metal finishing processes commonly, aqueous- based or vapour phase solvent media are used which can Effectively remove relatively weakly bound organic materials. The aqueous-based media are often combined with agitation including ultrasonic's for maximum speeding efficiency.

An Experimental Study of Water Jet Cleaning Process

Very high speed water jets (80–200 m/s) are typically used in such cleaning operations. These Jets diffuse in the surrounding atmosphere by the process of air entrainment and this contributes to the spreading of the jet and subsequent decay of pressure. Estimation of this pressure decay and subsequent Placement of the cleaning object is of paramount importance in manufacturing and material processing Industries. Also, the pressure distribution on the cleaning surface needs to be assessed in order to understand and optimize the material removal process. High speed water jets in air are extensively used in manufacturing industry for cutting and cleaning operations. Water jets are used for removal of various coatings or deposits from the substrates and also for the cutting of many materials. While water jet cutting involves the penetration of a solid material by a continuous jet, water jet cleaning involves an erosion process by which deposits are removed from the material surface. The figure below represents the anatomy of high speed water jets in air. There are the three main regions presents in water jet. The figure shows the high speed water jet used for cleaning purpose.

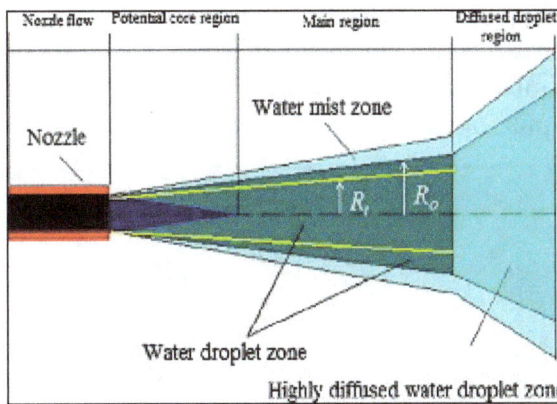

Anatomy of high speed water jets in air.

- Potential core region: This region is the one close to the nozzle exit. In this region, primary as well as secondary Kelvin–Helmholtz instabilities bring about transfer of mass and momentum between air and water. The process of air entrainment breaks up continuous water into droplets. There remains a wedge shaped potential core surrounded by amixing layer in which the velocity is equal to the nozzle exit velocity.

- Main region: The continuous interaction of water with surrounding air results in the break up of the water jet stream into droplets; the size of which decreases with the increase of radial distance from the axis. Since the jet transfers momentum to the surrounding air, its mean velocity decreases and therefore it expands. The region closest to the jet-axis is known as the water droplet zone. There is another zone; the water mist zone, which separates the droplet zone from the surrounding air. This mist zone is characterized by drops of very small size and negligible velocity.

- Diffused droplet region: This zone is produced by the complete disintegration of the jet into very small droplets having negligible velocity.

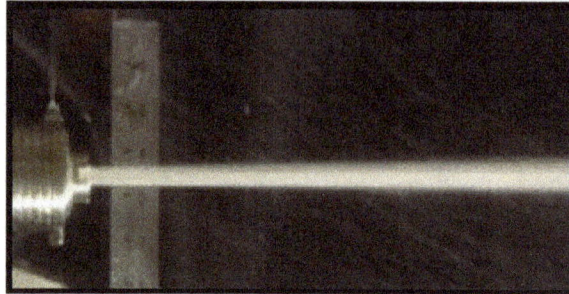

High speed water jet for cleaning operations.

Surface Finishes for Casting Processes after Cleaning

According to the data, gathered by Product Development and Analysis (PDA), Naperville, Ill., plaster casting produces the best surface finishes on average, with average root mean squared (RMS) values between 40 and 125. Other ceramic based processes, including investment casting, are similarly effective at delivering the high gloss shine. When the metal caster at these facilities pulls out all the stops, he or she can deliver RMS values as low as 25.

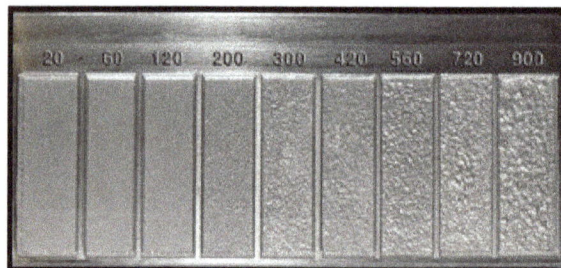

American Foundry Society's C-9 Micro finish.

Comparator Shows Surface Finishes from 20 to 900

RMS

Also at the low end of the spectrum, particularly when additional measures are taken, are the metal molding processes. Die casting and squeeze casting both can reach 20 RMS in extreme cases. Sand processes typically produce the roughest surfaces, but chemically bonded sands, including shell, nobake and cold box, can rival the ceramic processes at their best. Shell sand produces the most favorable range of average RMS values at 75-150, followed closely by vacuum casting (150-200 RMS). The figure shows the surface finish RMS values by AFS.

Millipore

Millipore is Italian terminology defined as maximum value of the impurities present in the component after cleaning the casting. It is a degree of cleanness. As per the requirement of the client it is very important to maintain the Millipore Value 16. This accuracy is not maintained by using conventional cleaning machine like shot blasting, abrasive jet machine, shot Peener because, these techniques cannot remove core particles that are present inside casting body. Equipments used in Millipore testing are as follows:

- Weighing Machine.
- Bicker with filter paper holding stand.
- Half H.P. suction motor.

Millipore Procedure

- First clean the component for which Millipore value is to be calculated.
- For cleaning purpose we use CTC (carbon tetra chloride) or Isopropyl alcohol.
- Collect the impurities in the tray.
- Take 5 micron filter paper & weigh (W1) it.
- Then hold the paper in the holder & filter the impurities through it.
- Then weigh (W2) the paper again.

The difference (W2-W1) gives the Millipore value.

Inspection and Testing of Castings

Two basic objectives of inspection are (i) to reject castings that fail to meet the customer's requirements, and (ii) to serve as a means of maintaining the quality of workmanship and materials used in the foundry. Inspection of castings broadly covers a large number of methods and techniques used to check the quality of castings. These methods may be classified into five categories:

- Visual inspection;
- Dimensional inspection;
- Mechanical and chemical testing;

- Flaw detection by non-destructive methods; and

- Metallurgical inspection.

Visual Inspection

All castings are subjected to a visual inspection to ensure that the surfaces fulfil the requirements of both the customer and the producer. Visible defects that can be detected provide a means for discovering errors in the pattern equipment or in the moulding and casting process. Most of the defects can be discerned by careful visual examination. Visual examination may prove inadequate only in the detection of sub-surface or internal defects in which case more sophisticated methods may be necessary.

Dimensional Inspection

Dimensional control is usually required for all types of castings. Sometimes it is not so critical but at other times it may be vital. When precision castings are produced by processes such as investment casting, shell moulding and die casting, dimensions need to be closely checked. Initially, when the castings are made from a new pattern, a few sample castings are first made which are carefully checked with the drawings to ensure that the sizes obtained conform to those specified and will be maintained within the prescribed tolerances in the lot under production. On testing of the sample lot, deviations from the blueprint are rectified on the pattern equipment. When the castings are found to be consistently within the tolerances, spot checks, together with a regular check of the patterns and dies being used, may be sufficient. In the case of the jobbing type of foundry, each casting produced may be different and, therefore, according to the customer's requirements, each one may have to be thoroughly inspected for dimensional variations.

Dimensional inspection of castings may be conducted by various methods:

- Standard Measuring Instruments: To Check the Sizes Instruments such as rule, vernier calipers, vernier height gauge, vernier depth gauge, micrometers, scribing block, combination set, straight edge, squares, spirit level, and dial indicator are commonly used. For high precision castings or after machining, more advanced measuring instruments, such as auto-collimator, comparator, ultrasonic instruments for measuring wall thickness and projection instruments are also required.

- Templates and Contour Gauges: For the Checking of Profiles, Curves, and Intricate Shapes Templates act as time-saving aids in measurement and facilitate the entire job. These can be easily prepared in mild steel or brass sheet by marking out, and cutting and finishing the profile that is required to be checked on the castings.

- Limit Gauges: For toleranced dimensions on casting produced on a repetitive basis, limit gauges are usually used. The type of limit gauges—plug, ring, snap, plate—depends on the shape of the parameter to be checked. Periodical checking and maintenance of limit gauges is very important.

- Special Fixtures: Special fixtures are required to be designed and used where dimensions cannot be conveniently checked by using instruments, for instance, during the checking of locations, relative dimensions, centre-to-centre distance, angularity of surfaces.

- Coordinate Measuring and Marking Machine (CMM): This machine is very useful for measurement and inspection of uneven, undulated, irregular, or curved surfaces which cannot be conveniently or accurately checked by other measuring tools or instruments. The accuracy of measurement of these machine ranges from 0.001 mm to 0.05 mm. Besides measuring, it can be used for marking purposes also in all three dimensions on metallic or non-metallic surfaces. Measurement and marking are accomplished easily without errors in reading in all three dimensions. Once the machine is set, all measurements can be carried out in a programmed sequence automatically. The machine in reality is a multi-axial device providing measurement of output of position and displacement sequentially without a need for changing tools.

The machine essentially consists of a touch probe, usually having a ruby tip, which is mounted on a horizontally sliding arm, movable vertically along a column. The column is fixed to a base which in turn is held on a large accurately machined granite surface plate and is movable in a direction perpendicular to the direction of the movement of arm. Thus, the probe is capable of being moved along all three axes for carrying out measurement of different surfaces of a workpiece. The sliding movements of the arm and column are performed with great precision and are read on an electronic digital read-out unit, attached to the machine. When marking is to be done on surfaces, a scriber is used in place of a probe. A larger variety of probes, scribers and other accessories are available to enable the machine to be highly flexible and accurate in operation. The movements along the three axes may be manual or motorised. The machine can be further equipped with a small computer system for processing the data obtained from measurement and for storing and retrieving the same.

A special software is also available with the computer so that measurement and inspection of different types of surfaces can be carried out automatically without the need for manual control. The drawing data from CAD station can be also transmitted to this machine by interlinking the two systems with the actual value of dimensions. A printer can also be provided with the computer for producing a hard copy of the inspection report. The CMM machines are now getting increasingly popular in inspection departments attached to tool rooms, pattern and die shops, foundry and forging shops, press shops, welding and structural shops and plastic and glass-part manufacturing units.

Mechanical and Chemical Testing

All foundries should have facilities for determining the mechanical properties of a cast metal and its chemical composition. Mechanical testing methods include certain procedures which require a standard type of equipment. These are:

- Tensile test to determine the tensile strength, yield strength, percentage elongation, and percentage reduction in area.

- Bend, notch bend, and impact transverse tests to evaluate the ductility and resistance to shock of the cast metal.

- Hardness test, which can indicate the strength and ductility of the metal (often, only hardness testing is conducted with visual inspection. Other tests are used only when so required).

- Fatigue test, applied in cases where an appraisal of the life of the casting in service is to be known.

- Tests for damping capacity and wear resistance.

Chemical testing is required to determine allowable limits. In the case of ferrous castings, it is necessary to know the percentage of carbon, silicon, sulphur, manganese, and phosphorus contents. The presence of alloying elements or metallic inclusions, such as Cr, Ni, Cu, Mg, W, V, Mo and Co, may also have to be determined. Chemical analysis can be used in all such cases to accurately ascertain the composition, though certain tests may be too cumbersome and time-consuming.

In many instances, it is necessary to quickly determine the content of carbon, silicon, and sulphur. This may also have to be regularly checked where a close control of the composition of metal is consistently required. Certain quick tests have been developed for such cases. One such test, commonly used in the case of grey iron, malleable iron, and ductile iron, is called Carbon Equivalent Measurement.

Carbon equivalent (CE) is given by:

$$\text{Total carbon \% } + \frac{1}{3} S \%$$

Since silicon has a predominant effect on the graphitising tendency of iron, the cumulative effect of carbon and silicon can indicate the strength characteristics of the iron produced. For shop floor use, a measuring instrument, known as 'instant carbon sensor' that quickly assesses CE. is available. It works on the principle that CE is directly related to the liquidus arrest temperature of the metal. The instrument is equipped to hold molten metal in a small reservoir over a chromel-alumel thermocouple and the temperature of the metal, as it solidifies in the reservoir, is registered on the chart of the temperature recorder. From this chart the liquidus arrest temperature can be easily determined. A conversion table is available below for arriving at CE corresponding to the determined value of liquidus temperature.

Similarly, for rapid determination of silicon in cast iron, a special apparatus called the Strohlein thermoelectrometer is available. It is equipped to measure the thermal emf produced when a junction between the metal whose silicon is to be determined and another metal, such as copper, is heated. The thermal effect depends on the actual metals involved and on the temperature difference between the junction and the cold ends. Silicon possesses a special position in the thermoelectric series and it is possible to determine its content by means of empirical calibration curves. The equipment is so designed that constant operating conditions are maintained at all times. Calibration curves are first prepared by using samples of known silicon content. The metal under test is then taken in the form of shavings and kept between two anvils on the apparatus. One of the anvils is heated by circulating a thermostatic liquid through it and the thermal emf generated due to the heating of a dissimilar junction is measured on a transistorised micro-voltmeter. From the calibration curves, the percentage of silicon is determined from the emf value.

Table: Conversion table for liquids temperature to carbon equivalent.

Liquidus Temperature	Carbon Equivalent
2250	3.60
2230	3.70
2210	3.80
2190	3.90
2170	4.00
2150	4.10
2130	4.20
2110	4.30

Chemical analysis, though the most accurate and reliable method, takes a long time. When metal is melted and refined in an electric furnace, the composition needs to be quickly determined, so that alloys can be added to adjust the constituents to the desired proportions. In such cases, chemical analysis is not suitable.

More rapid methods are available not only for CE or silicon, as mentioned earlier, but also for most of the other elements, which may be present in the metal even in traces. These methods are based on the principles of spectroscopy. Spectroscopic analysis is gaining popularity in foundries for quick determination of the constituent elements including the trace elements. Various types of spectroscopic analyzers are available, the selection depending on the nature of requirements in terms of the elements to be checked, the accuracy desired, the frequency at which tests are to be conducted, and the type of cast metal.

A microprocessor-based system operating on the principle of thermal analysis is also available for quick determination of carbon equivalent, total carbon, silicon and temperature of molten metal. It can be used for various cast metals like grey iron, malleable iron, SG iron, steel and copper-base alloys. The instrument is equipped with a digital display. A print-out is also obtained for permanent record. It has a high degree of accuracy, e.g., within ± 0.05% for carbon equivalent and total carbon and within ± 0.15% for silicon.

Flaw Detection by Non-destructive Methods

Non-destructive tests are also required to be conducted in foundries to examine the castings for any sub-surface or internal defects, surface defects, which cannot be detected by visual examination and for overall soundness or pressure tightness, which may be required in service. These tests are valuable not only in detecting but even in locating the casting defects present in the interior of the casting, which could impair the performance of the machine member when placed in service. Parts may also be examined in service, permitting their replacement before the actual failure or breakdown occurs.

The important non-destructive test for castings include:

- Sound or percussion test (stethoscope test),

- Impact test,

- Pressure test,

- Radiographic examination,

- Magnetic particle inspection,

- Electrical conductivity test,

- Fluorescent dye-penetrant inspection,

- Ultrasonic test, and

- Eddy current test.

1. Sound or Percussion Test (Stethoscope Test): This is an old method, which has been refined over the years. Basically, it entails suitably supporting the suspension of the part by chains or other equipment, permitting the part to swing free of the floor and other obstructions, and then tapping it with a hammer. The weight of the hammer blows is so adjusted that vibrations will be set up in the casting producing a certain characteristic tone which may or may not change the wavelength of sound produced by the blows.

The stethoscope test serves to detect relatively large discontinuities in an otherwise homogeneous metal and may be successful when applied to simple shapes and uniform cross sections. The drawback of the method is that it is difficult to judge the extent of the defect and to locate the fault.

2. Impact Test: This test may be destructive or non-destructive in nature, depending on the quality of casting. Moreover, it cannot be used in all cases as it can damage the casting.

A hammer of appropriate size is used to strike or fall on certain members of the casting where the defect is suspected. It is expected that the casting containing harmful defects will break and will thus be automatically rejected whereas those that are faultless will stand the test.

A variation of this method involves dropping the casting from a specified height onto a steel base. Obviously, the method of testing is not very reliable and sometimes even the defect-free castings may break. This method is therefore sparingly used these days.

3. Pressure Test: This method is employed to locate leaks and to test overall strength of certain parts, such as cylinders, valves, pipes, and fittings, which are required to hold or carry fluids in service under various amounts of pressure. The fluid used in testing may be water, air, or steam. Water being incompressible is generally preferred since danger is minimized even if the casting should shatter due to pressure. The pressure may vary from one and a half times to two times the working pressure. For safety reasons, the pressure is generally applied by means of a small hand pump. A leak, even if it is not located immediately, may be detected on the pressure gauge. Steam tests have the advantage that steam can press through smaller holes or openings through which water may not readily pass. Besides, the heat of the steam also causes minute cracks to widen due to expansion. While testing pneumatically, the casting is immersed in a tank carrying water and then the air pressure is applied. If there is a leak, air bubbles are formed.

4. Radiographic Examination: Radiography is a non-destructive test for detection of internal voids in castings. Electromagnetic waves having low wavelengths (varying between 10-6 and 10~10 cm) are used as a means of inspection. These waves, generally called X-rays, have properties similar to those of light waves, but they have much shorter wavelengths, which lie outside the range of human sensitivity. These X-rays can, however, be detected by a sensitive photographic film. Owing to their shorter wavelengths, these waves can penetrate materials that are normally opaque to light. The denser the material, the shorter the wavelength required to penetrate it. The test can be applied to all grades of iron and steel castings, though it is an expensive method of inspection.

The X-rays are produced by an X-ray tube which carries two sealed copper elements, the cathode and the anode. The cathode bears an electrically heated filament which generates electrons; when these electrons strike the tungsten target fixed to the anode they are driven towards the positively charged anode. The striking of the electrons causes their kinetic energy to be partly converted into heat, which is conducted away through the cooling fins provided on the anode and the remainder of the energy is converted into electromagnetic waves, termed X-rays. The X-rays pass out of the tube through a window in the form of a beam. The intensity of these X-rays is controlled by regulating the current passing through the filament. Similarly, the wavelength of the ray is inversely proportional to the voltage applied between the two poles. The shorter the wavelength, the greater the depth of penetration.

If there is a cavity or a hole in the casting under inspection, and, when such a casting is kept against the X-rays, the rays finding less obstruction penetrate more freely than at the place where the metal is more dense and solid. The rays that penetrate and emerge from the casting are absorbed by a photographic plate. Thus the part of the photographic plate opposite the defect will receive more rays and will be more exposed than the rest of the plate. This will produce a contrasting image on the negative. For more accurate results, special films with an emulsion coating are found suitable. Sometimes, in place of a photographic plate, fluoroscopic screens are used; these screens are made of materials that fluoresce when exposed to X-rays in a dark room. To protect the viewer from continuous exposure to the rays, the image of the screen is observed in a mirror, which is so placed that observer is located out of the path of the X-rays. The voltages required for the X-ray machines depend on the density of the metal and its section thickness.

Like X-rays, gamma rays which are emitted during the decomposition of radium, are also suitable for the inspection of castings. The wavelengths of these rays range between 10~75 and 10~105 cm and since these are shorter than X-rays they can penetrate metals more easily. Due to their high penetrating power, the radiation absorbed by the photographic film is negligible and the remainder passes through the film. Further, the difficulty experienced during the observation of thick and thin sections simultaneously is also less less than when using X-rays. But due to the high cost of radium and the need for expensive protective equipment, the technique is used to a limited extent.

Radiography does not enable detection of cracks. The position of defects in the section also cannot be easily defined unless special techniques are employed. Interpretation of radiographs depends on a subjective assessment and hence requires proficiency in the work along with experience. Castings which have passed the radiography test may still not be entirely leak-proof. Recommended radiations and their sources are given in table:

Recommended radiations and their sources			
	Radiation	Iron thickness	Exposure
X-rays	100 kV	< 12	
	200 kV	12-40	1-10 min
	400 kV	40-90	
	1000 kV	50-150	
	20,000 kV	60-250	
Y-rays	Cobalt60	40-100	
	Iridium 192	12-100	3-6 h
	Cesium 137	20-200	

5. Magnetic Particle Inspection: This test is used to reveal the location of cracks that extend to the surface of iron or steel castings, which are magnetic in nature. The casting is first magnetised and then iron particles are sprinkled all over the path of the magnetic field. The particles align themselves in the direction of the lines of force. Their distribution is also in proportion to the strength of the magnetic field. In the case of a faultless casting, particles will be distributed uniformly all over the surface, whereas if a defect exists, the iron particles will jumble round the defect. The reason is that a discontinuity in the casting causes the lines of force to bypass the discontinuity and to concentrate around the extremities of the defect. By studying the concentration of the particles, the depth at which the defect occurs can also be judged. However, considerable experience is necessary for an accurate estimation of the defects. With correct test procedure, cracks longer than 1 mm and a fraction of a millimetre deep can be detected.

Generally, a casting can be magnetised by passing an electric current through it. The current may be either alternating or direct. An alternating current is used when high surface sensitivity is desired, and the direct current is preferred where defects are to be located beneath the surface. Other methods for magnetizing castings include positioning the casting between two magnetic poles or placing the casting in a coil carrying a direct current.

Iron particles may be applied either dry with a handshaker or bulb blower or in wet form by spraying or pouring over the surface. When wet, the particles are carried in suspension form in liquid, for instance, kerosene, gasoline, or carbon tetrachloride. After testing, casting remains magnetised unless subjected to demagnetisation.

6. Electrical Conductivity Method: In this method, current is passed through the casting and read on an ammeter. If the casting has imperfections, there is a resistance to the flow of current and this is evident by a drop in the reading. The method is difficult to apply in practice owing to variations in sectional thickness, size and metallurgical structure; also it cannot be used directly unless a suitable standard is developed for a given lot of castings.

7. Fluorescent Dye-Penetrant Inspection: Penetrant testing helps to direct small surface cracks in castings, which cannot be observed with the naked eye. Although this method shows up the finest surface defects in a magnified form, interior defects, where the penetrant does not reach, cannot be revealed.

The method is very simple and can be applied to all cast metals. It entails applying a thin penetrating oil-base dye to the surface of the casting and allowing it to stand for some time so that the oil passes into the cracks by means of capillary action. The oil is then thoroughly wiped and cleaned from the surface. If the casting under inspection has any surface cracks, the oil will remain in these cracks and will tend to seep out. To detect the defects, the casting is painted with a coat of whitewash or powdered with talc and then viewed under ultraviolet light. The oil, being fluorescent in nature, can be easily detected under this light, and thus the defects are clearly revealed. By close observation of the amount of penetrant coming to the surface, the form and size of the crack can also be estimated with a fair degree of accuracy. The oils used are water-emulsifiable penetrants and are of proprietary nature.

Fluorescent dye inspection can also disclose those surface defects that are not revealed by radiographic inspection. For this reason, the penetrant test is often used to supplement the radiographic test.

8. Ultrasonic Testing: Ultrasonic testing used for detecting internal voids in castings, is based on the principle of reflection of high-frequency sound waves. If the surface under test contains some defect, the high-frequency sound wave, when emitted through the section of the casting, will be reflected from the surface of the defect and return in a shorter period of time. On the other hand, if the section is homogeneous and faultless, the wave will be reflected back after it travels through the whole of the section. In this case, it will take longer to return to its source. For detecting the lengths of time, an oscillograph is used. The path of travel of sound wave is plotted on the CRT screen of the oscillograph where it can also be measured. The advantage this method of testing has over other methods is that the defect, even if in the interior, is not only detected and located accurately, but its dimension can also be quickly measured without in any way damaging or destroying the casting. With a clean metal of small grain size, holes as small as 0.025 mm in diameter can be detected.

Ultrasonic testing can be applied to spheroidal graphite, compacted graphite, malleable and high-grade iron castings. The ease of application depends on casting shape. Proper test procedure can detect almost any hole, cavity or discontinuity. The method can be adopted for measuring wall thicknesses when only one side is accessible. Defects closer than 2 mm to a surface need special techniques to be detected.

The test frequencies used for detection vary from 0.5 to 5.0 MHz, according to the nature of iron and section thickness. For example, for SG iron having 20-50 nun section thickness, 5 MHz frequency is required, whereas for Grade 20 grey iron, 1 to 1.5 MHz is adequate. The equipment is light and portable; weighing 5 to 6 kg and can be taken anywhere at the work sites.

9. Eddy Current Test: This test is used for rapidly checking the hardness of iron castings. In this method of non-destructive testing, a coil carrying alternating current induces an eddy current of the same frequency in the test part under investigation. The eddy currents produced are affected by changes in the electrical conductivity, magnetic permeability and physical and metallurgical properties of the test part.

The instrument consists of: (i) a main unit, with a cathode ray tube (CRT) video display complete with frequency selector, oscilloscope controls, coil balance and sensitivity controls, and phase-shifting controls, and (ii) a matched pair of coils. The physical or metallurgical characteristics of any two parts kept in these coils are electromagnetically compared by observing their signals on the CRT screen. Before actual testing, the instrument has to be balanced. For this purpose, two similar parts are kept in the two coils and the test frequency is adjusted to the optimum value suiting the parameters of the test. The instrument is then balanced to obtain a horizontal straight line on the screen, showing that both parts are identical. Calibration curves are prepared and used for regular inspection.

One of the two parts kept originally for balancing is replaced by the part under test. The dissimilarity in shape of the signals, as observed on the CRT screen, indicates the variance in the concerned property of the two parts.

The eddy current intensity is greatest at the surface of the specimen and decreases as the depth increases. At high frequencies, eddy currents are produced only in the skin region, enabling the study of case depth, case hardness, and surface flaws or imperfections. Use of lower frequencies can be made to study sub-surface flaws, segregation, grain structure and chemical composition. The depth of penetration also depends on the conductivity and relative permeability of the specimen. Thus, the optimum frequency suiting the conductivity, magnetic permeability and depth of penetration desired must be selected such that the signals are clear and easily interpretable.

The eddy current method is suitable for testing hardness of rolled, forged, extruded, sintered or cast components in ferrous and nonferrous alloys, particularly, heat treated castings, such as malleable iron, nodular iron and CG iron. Hardness can be predicted to ±10 points Brinell. The test is normally limited to castings which will fit inside a 300-mm dia. coil. It is particularly well-suited for inspection of components produced by mass production machines on the shop floor, where the components produced have to be simultaneously inspected and segregated into good or bad lots or into different categories according to the quality.

Metallurgical Inspection

Metallurgical inspection is very useful for checking grain size, non-metallic inclusions, sub-microscopic pin holes, the type and distribution of phases present in the cast structure, and the response to heat treatment. These features can be appraised by certain methods:

- Chill test;
- Fracture test;
- Macro-etching test;
- Sulphur print test; and
- Microscopic examination.

1. Chill Test: Wedge test is a common method for chill testing of grey iron. It offers a convenient means for an approximate evaluation of the graphitising tendency of

the iron produced and forms an important and quick shop floor test for ascertaining whether this iron will be of the class desired. The depth of chill obtained on a test piece is affected by the carbon and silicon present and can therefore be related to the carbon equivalent, whose value, in turn, determines the grade of iron.

In practice, a wedge-test specimen of standard dimensions (IS: 5699-1970) is cast in a resin or oil-bonded sand mould. The test specimen is removed from the mould as soon as it is completely solid, quenched in water and then fractured in the middle by striking with a hammer. The chilled iron at the apex of the wedge usually consists of two zones; the portion nearest the apex entirely free of graphite is 'clear chill' followed by the portion in which spots of cementite or white iron are visible, called 'mottled zone'. The width of the chilled zone, measured parallel to the base and across the wedge is designated as 'total chill'. The value should not exceed more than half the value of the base. Chill width is largely affected by the use of alloy additions or inoculants and therefore the same value should not be expected in all cases.

Standard wedge Number:

Wedge No.	Breadth mm	Height mm	Included angle dec.	Length mm
1	5	25	11.5	100
2	10	30	18	100
3	20	40	28	100
4	25	45	32	125
5	30	50	34.5	150

2. Fracture Test: By examining a fractured surface of the casting, it is possible to observe coarse graphite, mottled graphite or chilled portion and also shrinkage cavity, pin hole, etc., The apparent soundness of the casting can thus be judged by seeing the fracture.

In case of steel casting, fracture test is also used in some foundries to quickly judge the amount of carbon present. A test rod of 25-mm diameter and 75-mm length is cast in a sand mould and quenched in water. The rod is then broken into two pieces and the fracture examined visually. Due to quenching, martensite is formed which is seen in the fractured section in the form of white spots or lines according to the amount of carbon present.

3. Macro-etching Test (Macroscopic Examination): The macroscopic inspection is widely used as a routine control test in steel production because it affords a convenient and effective means of determining internal defects in the metal. Macro-etching may reveal one of the following conditions:

- Crystalline heterogeneity, depending on solidification;

- Chemical heterogeneity, depending on the impurities present or localised segregation;

- Mechanical heterogeneity, depending on strain introduced on the metal, if any.

The test entails etching the sample piece of casting in a suitable reagent at a particular temperature for a prescribed length of time. The heterogeneity in the metal is revealed by the difference in chemical relations between the structural components of the metal and the selected etching reagent. Surface defects, inclusions, segregated area, etc., are selectively attacked by the reagent, and are therefore easily detected. Macro-etching reagents found suitable for steel and cast iron include hydrochloric acid, nitric acid, and Stead's reagent.

4. Sulphur Print Test: Sulphur may exist in iron or steel in one of two forms: either as iron sulphide or manganese sulphide. The distribution of sulphur inclusions can be easily examined by this test. The component to be examined for sulphur segregation is sectioned, ground, and polished. A sheet of photographic bromide paper is soaked in 2% solution of sulphuric acid for about five minutes. It is then removed from this acid solution and allowed to drain free from excess solution or is lightly pressed between two pieces of blotting paper. The emulsion side of the paper is then placed on the polished surface of the sample under moderate pressure for about two minutes. Care should be taken to ensure that no air bubbles are trapped. The paper is then removed and found to have brown stains where it was in contact with any sulphides. The reaction of sulphuric acid with the sulphide region of the steel produces H2S gas, which reacts with the silver bromide in the paper emulsion, forming a characteristic brownish deposit of silver sulphide. The darker and the more numerous the markings, the more the sulphur indicated. The paper is finally placed in a fixing solution for ten minutes, washed in running water, and dried. The entire operation can be carried out in daylight.

5. Microscopic Examination: Microscopic examination can enable the study of the microstructure of the metal or alloy, elucidating its composition, the type and nature of any treatment given to it, and its mechanical properties. In the case of all cast metals, particularly steels, cast iron, malleable iron, and SG iron, microstructure examination is essential for assessing metallurgical structure and composition.

The sample for examination is first cut to about 12-mm diameter and 9-mm thickness, and filed and ground to erase any deep grooves or marks. The piece should not get overheated at any time as this may alter its structure. The specimen is then polished on a series of emery papers of various grit sizes, the last one being of the finest variety. Sample polishing machines are available for the purpose. It may sometimes be desired to mount the sample in Bakelite, epoxy resin, or some plastic material before it is polished so as to keep edges from getting rounded off. For final polishing, the specimen is rubbed on a special cloth, which has already been impregnated with a polishing medium. It is then thoroughly cleaned and degreased, by washing in hot water, and sprayed with acetone or spirit.

The next step is to etch the specimen so that the etching reagent will first dissolve the thin bright layer produced during polishing and then attack metal at the grain boundaries and make them prominent on the surface. Owing to the nature of the grain boundaries, the rate of chemical solution along the boundaries will be greater than within the grains. Therefore, etching will produce the true underlying microstructure. The specimen is treated with the etching reagent for a few seconds until it acquires a dull matt appearance. It is then washed in hot water and dried in hot-air blast.

Etching Reagents

For steel and cast irons:

1. Nital. (2% solution of nitric acid in alcohol);

2. Picral. (4% solution of picric acid in alcohol); and

3. Alkaline sodium picrate (2g of picric acid and 25 g of caustic soda added to 100 ml of water).

For copper and its alloys:

- Ferric chloride solution in water or alcohol; and

- Ammonium hydroxide-hydrogen peroxide.

For aluminium and its alloys:

- Hydrofluoric acid solution in water (0.5%);

- Sodium hydroxide solution in water (1.0%); and

- Sulphuric acid.

After the specimen is etched and washed, it is ready for examination under the metallurgical microscope.

Scanning Electron Microscope

The use of a scanning electron microscope (SEM) has brought new insights in the field of metallurgical analysis, particularly in the study of fractures (fractography), grain size and grain growth, phase transformations, impurities and trace elements, characteristics of powders and their compaction. No specimen preparation is usually necessary. Even non-conducting materials can be examined by applying a mild conductive coating on the surface. The resolution of SEM being as high as and the depth of field being nearly 300 times that of an optical microscope, this instrument can be extremely valuable in quality control of castings, as also other products.

A fine beam of electrons is allowed to interract with the sample. The low-energy secondary electrons are made to strike a scintillator. The photon image is then fed to a photo multiplier through a light guide. The signal from the photomultiplier is used to influence the scanning in a cathode ray tube in synchronism with the scanning of the specimen by the original electron beam.

Heat Treatment

Heat treatment is an important step toward guaranteeing the mechanical properties of steel castings. Through molding, pouring, shakeout, and cleaning, castings take their final shape—but may not be strong enough or elastic enough for their final use. By heating and cooling metal at different rates, a foundry can change its mechanical properties.

Crystallization and Metal Properties

When molten metal cools, it freezes in crystalline structures. Under a microscope, these structures look like the frost crystals that form on glass in wintertime. Each structure grows from a center point until it meets with another crystal structure. These structures make up the "grains" of a metal.

The crystallization pattern of metal helps create its mechanical properties.

Just as varying winter conditions create lots of types of frost patterns, so variable temperatures change the crystals that make metal. The grain they create is usually invisible but are revealed when the metal is acid-etched.

The shape and relationship of the grains in an alloy determine its mechanical properties. Round grains can slide past each other when the metal is struck, denting rather than staying strong or breaking. Flat grains may stack together and support each other like bricks in a wall; stronger than the round grains, but still somewhat moveable. Jagged, interlocked grains may not have any give at all. The heat treatment of a metal can reshape its crystallization, which changes its grain, and therefore the metal's properties.

Work-hardening Metal

The image of a smith at his forge, pounding a glowing slab of metal, is immediately recognizable even though it is not a common sight any more. However, for much of human history, smiths would work metal mechanically to make it stronger. Today, rather than being hand-worked by a blacksmith, steel is often rolled to mechanically harden it.

Picturing the grain structure explains how work-hardening functions. Round grains within the metal are deformed, and their new shape gives the metal strength. In cold-rolling, for example, the round grains are squished and stretched to become more rod-like. These rods support each other, like sticks in a bundle. A smith or metalworker can hammer, twist, heat, cool, and stretch an object to change the shape of the grain. If the grains have no place to go when struck, they form an immovable, inelastic matrix which increases metal hardness.

However, this hardness can come with a cost: strength may render the material brittle. Irregularly shaped grains do not easily slide past each other: they are wedged together. Any sufficiently large impact—something greater than the strength of the bonds between grains—will break them apart.

Heat treatment like annealing creates rounder grains that create a more ductile metal.

Heat-treating Metal

The foundry starts creating desired mechanical properties of steel by choosing an alloy that is known to produce those characteristics. Yet there is very little control over the crystallization of this metal as the casting cools. Because crystallization creates the mechanical properties of the metal, the alloy may not behave optimally unless it is treated further. The foundry can do this by heating and cooling the metal in a controlled, regular fashion.

Heat treating is a non-destructive way to change material properties. It is sometimes a secondary process with work-hardened metal—but is the foundry's first choice, since the casting is already the right shape and can't be worked.

Crystallization almost always starts from the outer surfaces and moves in, and—especially in large castings—there is a large temperature differential between the shell of the casting and the center. The crystals grow irregularly, usually sharper and less malleable near the surface. They are often rounder and therefor softer the farther in. The casting shape and defects or inclusions within the metal will affect the rates of cooling, leading to zones in the metal which have dissimilar mechanical properties. These differences can cause internal metal strain, which can cause metal fatigue or failure. Heat treatment allows the foundry to go back inside a metal and rearrange the crystals that comprise it.

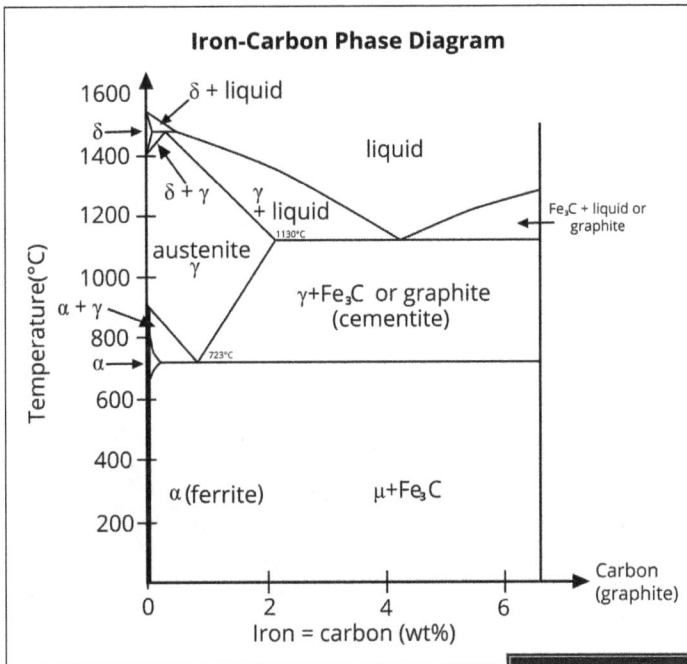

Iron-Carbon Phase Diagram.

Soaking

Soaking is the process that forms the foundation for all the heat treatment methods. Heat treatment relies on a metal's "recrystallization" temperature which sits below its melting point. During recrystallization carbon is unlocked to diffuse through the metal, moving from one molecular form to another depending on heat, carbon percentage, and time. This movement of carbon changes the crystallization patterns of the metal, and therefore carries different material properties. The iron-carbon phase diagram shows the formation of austenite, ferrite, pearlite, and cementite grains at different times and temperatures in the heat. Martensite, another grain structure found in hardened steels, is formed by cold-shocking austenite.

Soaking is therefore the process of bringing a casting above the point of recrystallization. The soaking "time at temperature" specified for a heat treatment allows the crystals in the metal to melt and reform. Looking at the iron-carbon phase cycle can help a foundry know how long to hold a casting at temperature to allow specific diffusion of carbon.

In most (but not all) parts of the iron-carbon phase cycle, soaking a cast or worked metal will make it less hard and brittle. As the grains in the metal grow more regularly, they are rounder and can rearrange on impact by sliding past one another. Also, since the item achieves the same temperature throughout, the crystals are usually more uniform than those in a freshly-poured casting.

Annealing

Annealing starts with soaking, and then continues by very slowly letting the steel cool in the furnace. The foundry worker turns the furnace off and allows a gentle, controlled drop in temperature. There is thermal consistency throughout the object both while heating and cooling, which means there are few internal stresses: no "zones" of metal with different crystallization properties occur. Metal that has been annealed is generally very malleable, with increased ductility, tensile strength, and elongation. The grain sizes with annealed metals are often very big due to the very slow cooling curve.

Normalizing

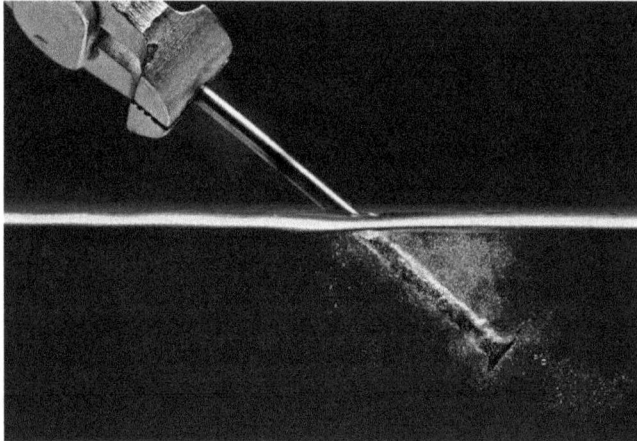

Quenching metals helps to harden them using thermal shock.

Normalizing a metal means bringing it up to recrystallization temperatures by soaking, and then pulling it from the furnace and allowing it to cool in the atmosphere. Many of the properties of annealed metals are evident in normalized metals, but because there is not quite the same evenness of cooling, the grains tend to be a little less regular. Still, a much smaller temperature differential than is found in freezing metal means a normalized product is less brittle.

The cooling rate found in normalizing creates smaller grains in the metal than annealing does, which means that in general, it will be stronger or harder than annealed metal.

Quenching

What if a very high degree of hardness is desired? When making tools and machine parts, softening the metal may defeat the purpose. Heat treating can allow hardness to be specified and consistent. To create hardness in steel, the foundry soaks the steel until austenite is the main molecule and then quenches it in cooler oil or forced air. When austenite is cold-shocked, it creates a slightly irregular crystalline structure called martensite. This material is harder due to a carbon distortion in each martensite molecule.

Since quenching happens from the outside-in, large objects can experience the pressures of rapid crystallization leading to internal pressure in the metal. These forces can sometimes cause cracking if the quench is too extreme. For this reason, water-quenching is not very common for large steel objects, as it causes a very rapid drop in temperature which can cause cracks to form. Oil and air both cool slightly less vigorously.

A retained fragment of sand core in an industrial impellor explodes during quenching.

However, it is not only steels that are quenched for hardening. Water quenching is used in a foundry. Non-steel metals may not suffer the same internal pressures because their phases and molecular structures will be different. Manganese is water-quenched at much higher temperatures than steel, without cracking. However, the difference in temperature is so great that any quenching deals with a lot of energy that can go awry. Below is an explosion caused by a retained sand core during the quenching of a manganese steel casting.

Tempering

Finding the right mixture of hardness-and-ductility can also be achieved through a process called tempering. Tempering is often done with quenched steel to make it less brittle while preserving some of the hardness. In tempering, a metal is reheated yet again, but now to a lower temperature than in annealing, normalizing, or quenching.

Martensite is not a stable molecule in heat—it is achieved under shock—so tempering steel means destabilizing the martensite to let it start to convert to cementite and ferrite. A range of temperatures and length of times in the temper oven will influence how much of the martensite is converted and therefore how soft the metal becomes. For example, metal springs may be tempered at higher temperatures for increased elasticity compared to tools tempered at lower temperatures to maintain hardness.

Tempering is often used to relieve the internal stresses in a quenched material. A metal that's undergone other heat stress like welding or blacksmithing can be tempered to allow the molecules within to relax a little into one another.

Variations in Heat Treatment

In a foundry, castings are usually uniformly heat treated. However, sometimes an item can be irregularly heat treated. Tempered steel swords commonly were variably tempered, such that the blades had hard edges while the cores stayed springy. Springs sometimes go through differential heat treatment, to match their function.

As with much of the foundry, an understanding of the chemistry of an alloy means that times, temperatures and tolerances can be scientifically specified. However, over time, a foundry worker comes to know the metal they're working with. Like an expert chef knows their ingredients well enough to not need a recipe, an expert foundry worker will know when something's off. A metal that takes too long to come to glowing, or cools down too fast, tells a molecular story to an experienced eye—without the help of equipment in the lab.

Machining Castings

Machining is used to introduce features that cannot be produced during the casting process. This is due to the very small tolerances of the design dimensions. In these instances, a 'machining allowance' is incorporated at the design stage so that the casting can eventually be machined back to the exact dimensions.

Machining takes place once any fettling or heat treatment has been completed but before any finishing processes, such as anodising or painting. Machining is carried out by computer numerical control (CNC). Specialist computers are programmed to guide the machining tools and shape the component accordingly.

Turning

In turning, carbide or ceramic cutting tools are used to create a smooth finish on the casting. These tools have good wearability, with long life and repeatability.

The component is positioned in a chucking (turning) fixture and rotated using either a vertical or horizontal machine. Using the tools, any unwanted material can be removed from the inside and outside of the casting to produce turned bores (holes) and diameters to close tolerances, with a high standard of finish.

Milling

In milling, the component is securely clamped to the machine table. The cutting tools, which use changeable carbide or ceramic inserts, move and rotate at optimum speed across the work piece to generate various features on the face of the casting. This can be carried out on a horizontal or vertical axis.

Traditionally, turning and milling have been carried out separately but we also have machines that can now perform both operations together – reducing set-up/tooling times and costs.

Surface Grinding

Surface grinding is used to produce a precision flat finish. The component is secured on a magnetic plate or, with non-ferrous castings, a holding fixture. The outside edges of the castings are ground using a carborundum grinding wheel, which revolves at a rapid speed to produce the required finish.

Manual Drilling

Manual drilling is used to drill or tap holes into the component. The positioning of the holes is determined using cast dimples which are introduced during casting manufacture or by bespoke drilling fixtures.

Inspection

The final stage of the process is designed to check that the machining work meets the specified criteria. Casting inspection is carried out using automated equipment and probes during the machining process and post manufacture.

Rough Machining for Metal Castings

Rough machining is important for exported metal castings. It can largely reduce the defective rates of castings, and can meet the dimensional requirements to some positions.

Rough machining is different with finish machining. After we complete the rough castings, or called as raw castings, we can do some coarse machining works, such as milling and lathe, and leave 0.5 to 1.0mm machining allowance for the further finish machining works.

Normally, rough machining works should be done by the metal foundries, and finish machining works were done by clients or entrusted professional machining workshops.

Why is Rough Machining Necessary?

Firstly, there are many types of casting defects beneath the surface of metal castings, such as shrinkage, cracks, pin air holes, sand inclusion etc. These defects can not be found during visual inspection. So, when the clients receive the casting products, and during machining process, these defects will be shown up, then it will cause more costs for return castings and machining costs. As for the exported casting products, retreating defective castings will cause much more costs. However, if metal foundries could do some rough machining in-house, they will be able to find these defective castings, and recycle the casting materials, so will reduce the loss for both themselves and their clients largely.

Secondly, as for some dimensions, rough casting cannot meet the accuracy requirements. If the required tolerance is out of rough casting tolerance range, then rough machining will be needed. Moreover, deformation is inevitable for rough castings, so rough machining could make the deformed surfaces more flat.

Painting and Finishing

Finishing is the process used to prepare cast components for industrial use.

Painting

Painting gives components exceptional wear and corrosion resistance and a good surface finish. The two main techniques used are powder coating and wet painting.

Powder Coating

Powder coating is one of the most popular finishing techniques that has been in use since the 1950s. It is employed on a whole host of products found all around us, from household goods, such as washing machines, through to aluminium windows, tractors and cranes.

Powder coating is a 'dry' process that produces a hard-wearing surface finish. The part is first thoroughly cleaned to remove any surface grease, oil, dust or other contaminants.

It can be pre-treated with phosphates or chromates to enhance the durability of the finishing coating.

Next, finely ground particles of powdered paint pigment are sprayed onto the surface of the component using an electrostatic spray gun. These particles are attracted to the part using the electrostatic charge. The part is then heated in a curing oven to melt the powder and form a hard, durable skin.

Powder coatings are available in a very large range of colours and they can produce varying textures, depending on the requirement and end use. These coatings offer higher levels of abrasion and wear resistance than wet paint finishes – due to the thermal bonding involved. They also produce thicker coatings than liquid paint and a more even surface finish.

Importantly, powder coatings offer many environmental benefits over liquid paints as they do not contain solvents and release little or no volatile organic compounds (VOCs) into the atmosphere. The materials can also be recycled. Although the initial set-up costs can be higher than wet paint techniques, powder coating offers long-term benefits with lower maintenance costs.

Wet Painting

Using this process, liquid paint is sprayed onto the surfaces of the casting in a coloured pigment suspended in a solvent. While it is being sprayed, the part is rotated to ensure full coverage. The casting is generally pre-treated with a primer to enhance the protective effects.

Wet spray painting offers several advantages over powder coating. It is particularly useful for parts that cannot be heated for powder coating. Furthermore, it can produce a thinner finish. Finally, wet painting is more economical for small production runs due to the relatively small set-up costs.

Surface Coating of Iron Castings

The most of iron castings need surface coating to prevent rust. According to the different functions, there are many types of surface coatings.

Primer Anti-rust Painting

In order to prevent the rust during shipping, normally, iron foundries will paint the anti-rust primer painting for iron castings. These paintings are very normal, so the prices are very low, it is about 30 USD/ton based on the total weight of castings.

Lead-free (Nonleaded) Anti-rust Painting

Normal primer painting has some content of lead, which will cause the pollution to the underground water. Therefore, some iron castings need lead-free painting. Lead-free painting is more expensive than normal primer painting, so they are not widely used except for the special uses.

RAL Painting

RAL painting means the painting lives up with the RAL standard and requirement to the colors. You need to buy these painting from professional painting suppliers. Therefore, their cost is higher than normal painting. RAL painting is for the finish painting, not for primer painting.

Zinc-rich Primers

Zinc-rich primer paintings have many advantages, such as quick drying, good physical properties, good anti-heat property and can be mixed to proper colors. Of course, zinc-rich primers are very expensive, it is about 150 USD/ton. So, it is not commonly used.

Polyurethane Coating

PU (Polyurethane) Coating has good durability and good surface flexibility. Its cost is about 150 USD/ton based on the normal casting weights. Not commonly used.

Electro Galvanizing

This kind of galvanize is also normally called as zinc plating. It is about 80 to 120 USD/ton according to the difference of surface areas. Electro galvanizing is commonly used for the high anti-rust requirements.

Hot Dip Galvanizing

Hot dip galvanizing is also called hot galvanizing, or called HDG, which is a kind of chemical galvanizing process. HDG has very good anti-rust properties, it can prevent the iron castings from rust for several years in the sea water. Since HDG has pollution and can consume more zinc, so its cost is high, it is about 230 USD/ton.

Heat Resistance Paint

Heat resistance paint is also called as fire-resisting paint, which is normally used for the stove parts. Since they need to resist hundreds of degrees heat, so they are more expensive than normal painting. It is about 220 USD/ton.

Asphalt Paint and Bituminous Paint

Bituminous paint is also called as asphalt paint, which is usually used for the pipe lines and fittings, or marine parts. Normal bituminous paint is not expensive, which is almost same as the normal anti-rust painting. However, many overseas clients will require the environmental bituminous paint. So, this kind of paint will be more expensive, it is about 60 to 70 USD/ton.

Besides, there are many other surface coatings for iron castings, such as electrophoresis coating, chrome-plating, powder coating, emulsion paint etc.

Pollution Control in Foundries

Pollutants in a Foundry

Foundries are among the industrial plants causing environmental pollution, producing substantial quantities of air pollutants. The numerous processes available for moulding, melting and casting are accompanied by evolution of heat, noise, dust and gases. Dust, fines, fly ash, oxides, etc., which form particulate matter are generated in large quantities when preparing mould and core sands and moulds, melting metals, pouring moulds, knocking out poured moulds and loading and unloading raw materials. Gaseous matter like gases, vapours, fumes and smoke are produced during melting and pouring operations. The major pollutants emitted from various work areas in a foundry are given in table. The basic means of controlling the emission of pollutants are changing the production process, sup plying adequate make-up air, proper aeration and ventilation of the shop, reduction of pollutants at source by taking appropriate control measures, dispersion and dilution of pollutants in the air space and good housekeeping.

Table: Major pollutants emitted in a foundry.

Work Area	Pollutant
Pattern shop	Sawdust, wood chips
Sand preparation	Dust and fumes,
	Powder materials
Moulding and core-making	Sand
	Fumes
	Binder dust
	Vapours
Mould drying and ladle heating	CO, SO_2
Cupola	SO_2
	CO
	Unburnt hydrocarbons
	Smoke

Cupola	Metallic oxides
	Coke dust
	Limestone dust, fly ash
Electric arc furnance	Dust, CO, So$_2$ oxides,
	Nitrogen cyanide, fluoride, etc.
Electric induction furance	Dust, oxides, smoke
Pouring and mould cooling	CO
	Binder fumes
	Oil vapours
Knock-out	Sand, fines and dust
	Smoke, steam, vapours
Fettling	Dust, metal dust, sand fumes
	Abrasive powder
Heat treatment	CO, SO$_2$, oil vapours

Dust and Fume Control

It is of utmost importance that the air polluted by foundry work be cleansed to maintain hygienic working conditions. The atmosphere in the pattern shop is charged with fine particles of sawdust. Dust sand particles are exuded when sand is mixed and prepared during moulding, shake-out fettling operations. Fumes are produced during melting, metal-transfer, and pouring operations. It is essential to devise a system for collecting all the dust and fumes so produced and disposing them so that they do not pollute the atmosphere in the foundry and pose a threat to the health of the workers. When a foundry layout is planned, provision should be made for dust and fume control. If this vital aspect is attended to as an afterthought, it becomes difficult to incorporate the necessary equipment.

Materials requiring to be separated may be classified into two broad categories: particulate matter, where the particles are either solid, such as dust, fume, smoke, and fly ash, or liquid, such as mist and fog; and gaseous matter, where the contaminant may be either gas over the entire range of atmospheric and process temperatures and pressures, or liquid at lower temperature, and gas at the temperature and pressure of its release into the atmosphere.

The method of separation depends on the category to which the pollutant belongs. Some separation processes are applicable to several types of pollutants whereas others to only one of them. The methods commonly used in foundries are now outlined.

Filter

The filter serves for removing particulate matter from gas or air streams by retaining it in or on the porous structure through which the gas flows. The porous structure is usually a woven or felted fabric. The filter must be continuously or periodically cleaned, or replaced.

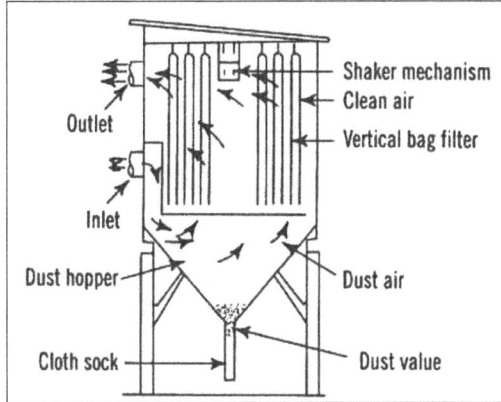

(A) Bag filter.

The filtering action may be obtained in various ways, such as direct interception, impaction, diffusion, and electrostatic precipitation. In direct interception, the particle is carried by a streamline of gas, which heads it directly towards a part of the solid surface comprising the filter. In impaction, the particle is in a streamline of gas, which sweeps by the solid material of the filter and allows the particle to touch the filter material. In diffusion, a blow from a molecule of the gas projects the particle to the filter surface. In electrostatic precipitation, electrical charges on the particle and filter attract the particle towards the filter. One or more high-intensity electrical fields are maintained to cause the particles to acquire an electrical charge and be forced to move towards the collecting surface.

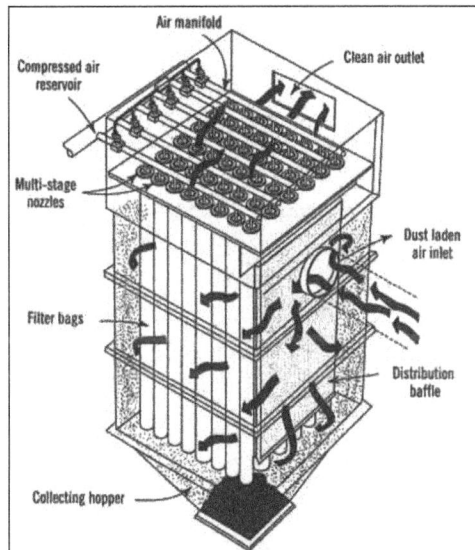

(B) Schematic arrangement of an ultra-jet type filter.

Filters are commonly employed in pattern shops on various woodworking machines, such as band saw, circular saw, and sanding machines. They are also used on cupola collection systems in conjunction with other equipment, such as after-burners, gas cooler, recuperators and exhaust blower. Sand-reclamation plants also use bag filters for separating 'fines' from sand grains.

Cyclone

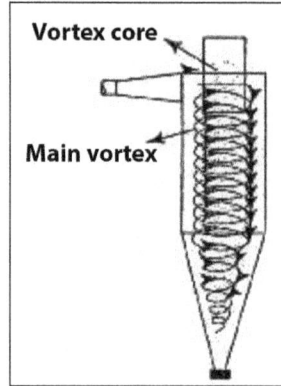

Cyclone.

The cyclone works on the principle of Vortex core centrifugal separation in which a vortex motion of the particulate matter is created within the collector. This motion provides the centrifugal force which propels the particles to locations from where they may be removed. Cyclones may be operated either dry or wet. Also, they may either deposit the particulate matter in a hopper or concentrate it into a stream of gas which flows to another separator for ultimate collection. The cyclone is used in sand-preparation plants for separating sand particles from air, in cleaning the cupola exhaust, in moulding shops, and on shake-out stations.

Mechanical Collectors

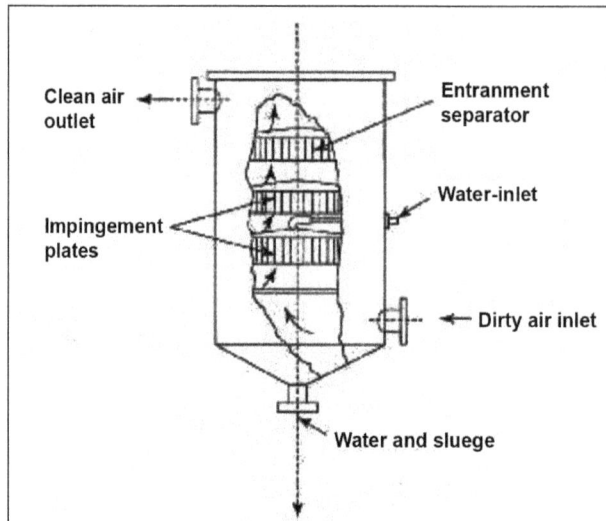

Wet centrifugal dust collector.

These devices include settling chambers, baffled chambers, and fan arrangement, which collect particulate matter by gravity or centrifugal force but do not depend upon a vortex as in the case of cyclones. As their efficiency of collection is generally rated low, they are used as precleaning devices before other types of collectors. They also function in combination with filters or scrubbers. Cupola exhaust systems often make use of mechanical collectors.

Scrubbers

The scrubber is employed primarily for removing gases and vapour-phase contaminants from the carrier gas, though it can also remove particulate matter. A liquid, usually water, is introduced into the collector and it either dissolves or chemically reacts with the contaminant collected. Methods used to effect a contact between scrubbing liquid and carrier gas includes: (i) spraying the liquid into chambers containing baffles, grille, or packing; (ii) flowing the liquid over weirs; and (iii) bubbling the gas through tanks or troughs of liquid. Scrubbers are ideal for cleaning the exhausts of cupola and arc furnaces.

After Burners

The after-burner assists in oxidising the solid combustible material present in the particulate matter and converts it into gaseous form. It also helps to convert carbon monoxide into carbon dioxide as in the case of cupola gases. After-burning may be accomplished by using furnace oil as a fuel and introducing it along with air into a combustion chamber through which the carrier gas passes.

Combination Devices

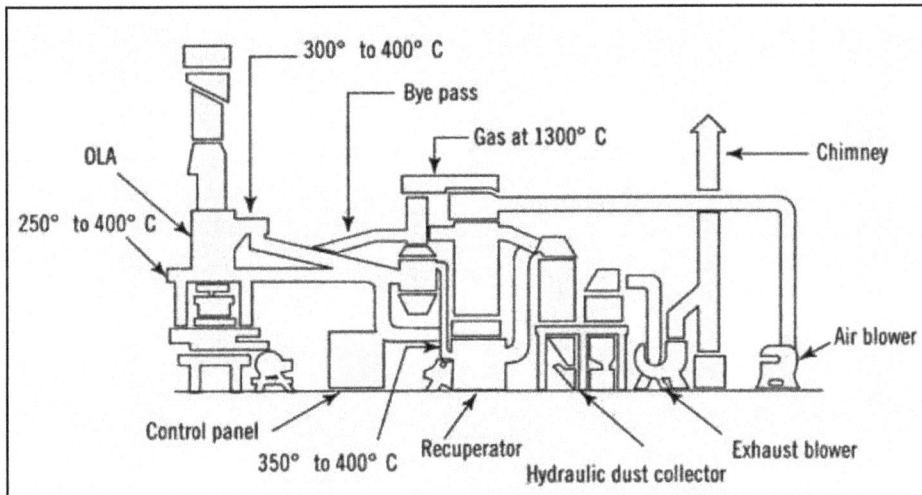

Cupola furnace installation.

Some devices combine features of the aforementioned equipment so that dust and fumes are controlled most economically and with a minimum pressure drop. For instance, there are cyclones in which liquid is sprayed, and scrubbers in which cyclonic action is used. Packed-bed filters, operated wet and packed-bed scrubbers are similar to each other, the only difference being that the equipment designed to separate particulate matter is called a filter and the same when designed to separate gaseous contaminants is called a scrubber. Often, equipment of different types are used in series. Figure shows a common arrangement of exhaust cleaning used on large-sized cupolas.

References

- The-casting-finishing-process, processes: vulcangroup.com, Retrieved 13 July, 2019

- Major-post-casting-finishing-processes-used-in-die-cast-components, die-cast-light-fixtures: pac-diecast.com, Retrieved 21 July, 2019

- Recent-trend-in-casting-cleaning-technique, paper: ijser.org, Retrieved 7 May, 2019

- Tempering-steel gref: reliance-foundry.com, Retrieved 18 April, 2019

- Machining-castings-an-overview, news: haworthcastings.co.uk, Retrieved 14 July, 2019

- Rough-machining-metal-castings: iron-foundry.com, Retrieved 11 January, 2019

- Painting-and-finishing, news: haworthcastings.co.uk, Retrieved 29 June, 2019

Permissions

All chapters in this book are published with permission under the Creative Commons Attribution Share Alike License or equivalent. Every chapter published in this book has been scrutinized by our experts. Their significance has been extensively debated. The topics covered herein carry significant information for a comprehensive understanding. They may even be implemented as practical applications or may be referred to as a beginning point for further studies.

We would like to thank the editorial team for lending their expertise to make the book truly unique. They have played a crucial role in the development of this book. Without their invaluable contributions this book wouldn't have been possible. They have made vital efforts to compile up to date information on the varied aspects of this subject to make this book a valuable addition to the collection of many professionals and students.

This book was conceptualized with the vision of imparting up-to-date and integrated information in this field. To ensure the same, a matchless editorial board was set up. Every individual on the board went through rigorous rounds of assessment to prove their worth. After which they invested a large part of their time researching and compiling the most relevant data for our readers.

The editorial board has been involved in producing this book since its inception. They have spent rigorous hours researching and exploring the diverse topics which have resulted in the successful publishing of this book. They have passed on their knowledge of decades through this book. To expedite this challenging task, the publisher supported the team at every step. A small team of assistant editors was also appointed to further simplify the editing procedure and attain best results for the readers.

Apart from the editorial board, the designing team has also invested a significant amount of their time in understanding the subject and creating the most relevant covers. They scrutinized every image to scout for the most suitable representation of the subject and create an appropriate cover for the book.

The publishing team has been an ardent support to the editorial, designing and production team. Their endless efforts to recruit the best for this project, has resulted in the accomplishment of this book. They are a veteran in the field of academics and their pool of knowledge is as vast as their experience in printing. Their expertise and guidance has proved useful at every step. Their uncompromising quality standards have made this book an exceptional effort. Their encouragement from time to time has been an inspiration for everyone.

The publisher and the editorial board hope that this book will prove to be a valuable piece of knowledge for students, practitioners and scholars across the globe.

Index

www.ingramcontent.com/pod-product-compliance
Lightning Source LLC
Chambersburg PA
CBHW061934190326
41458CB00009B/2737